T0200860

Effective Experimentation

Effective Experimentation

For Scientists and Technologists

Richard Boddy
Gordon Smith

Statistics for Industry, UK

A John Wiley and Sons, Ltd, Publication

This edition first published 2010
© 2010, John Wiley & Sons, Ltd

Registered office
John Wiley & Sons Ltd, The Atrium, Southern Gate, Chichester, West Sussex, PO19 8SQ, United Kingdom

For details of our global editorial offices, for customer services and for information about how to apply for permission to reuse the copyright material in this book please see our website at www.wiley.com.

Library of Congress Cataloguing-in-Publication Data

Boddy, Richard, 1939–
    Effective experimentation : for scientists and technologists / Richard Boddy, Gordon Laird Smith.
        p. cm.
    Includes index.
    ISBN 978-0-470-68460-3 (hardback)
    1. Science—Experiments—Statistical methods.   2. Technology—Experiments—Statistical methods.
I. Smith, Gordon (Gordon Laird)   II. Title.
    Q182.3.B635 2010
    507.2'7—dc22

                                                                                    2010010298

A catalogue record for this book is available from the British Library.

ISBN: 978-0-470-68460-3

Set in 10/12pt, Times Roman by Thomson Digital, Noida, India.
Printed and bound in the United Kingdom by Antony Rowe Ltd, Chippenham, Wiltshire.

Contents

Preface

This is a practical book for those engaged in research within industry. It is concerned with the design and analysis of experiments and covers a large repertoire of designs. But in the authors' experience this is not enough – the experiment must be effective. For this the researcher must bring his knowledge to the situations to which he needs to apply experimentation. For example, how can the design be formulated so that the conclusions are unbiased and the experimental results are as precise as needed?

Each chapter starts with a situation obtained from our experience or from that of our fellow lecturers. A design is then introduced and data analysed using an appropriate method. The chapter then finishes with a critique of the experiment – the good points and the limitations.

The book has been developed from the courses run by Statistics for Industry Limited for over 30 years, during which time more than 10,000 scientists and technologists have gained the knowledge and confidence to apply statistics to their own data. We hope that you will benefit similarly from our book. Every design in the book has been applied successfully.

The examples have been chosen from many industries – chemicals, plastics, oils, nuclear, food, drink, lighting, water and pharmaceuticals. We hope this indicates to you how widely statistics can be applied. It would be surprising if statistics could not be applied successfully by you to your work.

The book is supported by a number of specially designed computer programs and Excel Macros. These can be downloaded from Wiley's website. Although the reader can gain much by just reading the text, he/she will benefit even more by downloading the software and using it to carry out the problems given at the end of each chapter.

The book gives a brief overview of introductory statistics. For those who feel they need a more comprehensive view before tackling this book can refer to *Statistical Methods in Practice* (2009) by the same authors.

Statistics for Industry Limited was founded by Richard Boddy in 1977. He was joined by Gordon Smith as a Director in 1989. They have run a wide variety of courses worldwide, including Statistical Methods in Practice, Statistics for Analytical Chemists, Statistics for Microbiologists, Design of Experiments, Statistical Process Control, Statistics in Sensory Evaluation and Multivariate Analysis. This book is based on material from their Design of Experiments course.

Our courses and our course material have greatly benefited from the knowledge and experience of our lecturers: Derrick Chamberlain (ex ICI), Frits Quadt (ex Unilever), Martin Minett (MJM Consultants), Alan Moxon (ex Cadbury), Ian Peacock (ex ICI), Malcolm Tillotson (ex Huddersfield Polytechnic), Stan Townson (ex ICI), Sam Turner (ex Pedigree Petfoods) and Bob Woodward (ex ICI). In particular we would like to acknowledge Dave Hudson (ex Tioxide) who wrote the Visual-Basic-based software, John Henderson (ex Chemdal) who wrote the Excel-based software and Michelle Hughes who so painstakingly turned our notes into practical pages.

Supporting software is available on the book companion website www.wiley.com/go_effective.

Richard Boddy
Gordon Smith
Email: s4i@aol.com

April 2010

1

Why bother to design an experiment?

1.1 Introduction

There are many aspects involved in successful experimentation. This book concentrates mainly on designing and analysing experiments but there is much more required from you, the experimenter. You must research the subject well and include prior knowledge available from previous experiments within your organization. You should also consider a strategy for the investigation such as considering a series of small investigations. You must plan the experiment operationally so it can be successfully undertaken and, lastly, having analysed the experiment you must be able to interpret the analysis and draw valid conclusions.

If you follow that path, then you should have completed a successful project.

If not, then you may have wasted resources, had insufficient trials or data to be able to make conclusions that will stand up to scrutiny, or end up by making invalid claims.

There is no guarantee of finding all the answers, but you will have been well informed and will have made the most efficient use of the information and data available.

Let us consider some situations that illustrate the benefits of using designed experiments.

1.2 Examples and benefits

1.2.1 Develop a better product

An oil formulator has been charged with the task of improving the formulation of a lubricating oil in order to improve the fuel economy of motor engines. There are two important components of the formulation – type of base oil and level of friction modifier. Without knowledge of experimental design, he does not wish to change both variables at once. He keeps to the current level of friction modifier and changes the base oil, gaining an improvement. The next trial therefore uses the new base oil and he changes the level of friction modifier. It also

Effective Experimentation: For Scientists and Technologists   Richard Boddy and Gordon Smith
© 2010 John Wiley & Sons, Ltd

gains a small improvement. He reports that the new oil should be made with new base oil and changed level of friction modifier and can be called 'New Improved'.

This is an inefficient way of exploring the experimental space, even assuming that only two levels of each variable are possible. He has assumed that a change resulting from changing the level of friction modifier when using one base oil will be repeated with the other, but this does not often happen with manufacturing processes. There are often interactions. He should have tested all four combinations of the two levels of both variables in a factorial experiment. It may be that the best combination is none of those that he examined, as in the example in Chapter 4.

Such an experiment, if replicated (more than one trial at each set of conditions) would give him the following benefits:-

(i)   determination of the effects of each variable and knowledge of whether or not there is an interaction between them;

(ii) determination of whether an improvement can be made;

(iii) a measure of the batch-to-batch variability that enables him to test differences for significance.

1.2.2   Which antiperspirant is best?

A toiletries company has developed some formulations of an antiperspirant and wishes to determine which one is most effective. After chemical and microbiological tests the only realistic way is to test them out on volunteers in a carefully controlled environment. Perhaps at first thought a large number of volunteers should be assembled, and formulations allocated at random to the volunteers, each person testing one formulation. The trouble with this approach is that there is a lot of variation between one person and another in amounts of perspiration and the effectiveness of an antiperspirant, which would obscure any differences that there might be between formulations.

An experimental design is needed so that person-to-person differences ('nuisance' variation) can be identified but their effect removed when comparing the formulations. Thus, a panel of volunteers is gathered, and each one tests every formulation. The person-to-person variation would be there but would affect all the results, but differences between formulations should be more consistent. This design is known as a randomized block design, introduced in Chapter 18.

The benefits of this design are:-

(i)   formulations can be directly compared;

(ii) person-to-person variability can be quantified but its effect eliminated in the comparison of formulations;

(iii) the best formulation can be identified.

1.2.3   A complex project

Bungitallin Spices are developing a new spice for lightly flavoured cheeses. They have identified 30 ingredients, decided on a composition and produced a trial sample. The taste seems reasonable and they decide to proceed with a marketing campaign to launch their new product.

Now clearly Bungitallin have a great knowledge of spices and it is perhaps not surprising in the spice industry that 30 ingredients have been included. However, there are many questions that are readily brought to mind.

i. How did they decide on the composition?

ii. Could they have done better if they used experimental design?

iii. Do all 30 ingredients contribute to taste? How many can be discerned and at what levels? How many can be removed without any discernible effect in the taste?

iv. How many of the 30 ingredients are necessary for texture or other parameters and at what levels?

v. How do the ingredients interact with each other?

vi. How is the spice to be produced? What are the process conditions? How robust are the conditions?

vii. How much variation can we expect from batch-to-batch? Is this acceptable or does it need reducing?

Clearly there are a lot of questions to be answered. If we attempt to answer all the questions in an unstructured manner the cost may be far greater than the profit from launching the spice. On the other hand, if we do nothing Bungitallin may be left with a failure at great cost. Thus, we must investigate, but in an efficient way.

Experimental design offers an approach that will enable us to achieve our objective in an efficient manner and give us unbiased results, thus enabling us to have confidence in our conclusions.

Different chapters of this book will help you to answer these questions.

Questions iii), iv) and v) can be investigated using *factorial* or *fractional factorial designs* followed by *response surface methods* to achieve the best formulation. If the experiment is too large to carry out in one trial a *central composite design* may be employed.

Questions vi) and vii) can be investigated using *saturated designs* or *computer-aided experimental designs (CAED)*.

Question vii) can be investigated using *randomized block* or *Latin Square designs.*

1.3 Good design and good analysis

Of course, it is not only necessary to carry out a good design but it must be followed by a good analysis – in fact, when designing an experiment we should also consider how it is to be analysed.

This book starts with a chapter that covers *summary statistics, the normal distribution, confidence intervals* and *significance testing*. Later it refers to *multiple regression*, a necessary tool when the design has an imbalance which can occur for many reasons such as 'lost' data.

All these designs and methods of analysis will greatly enhance your experiments but we must not forget the most important aspect of experimental design – the researcher's knowledge. The design is aimed at increasing this knowledge and making it more rigorous so that we have a high degree of certainty that actions resulting from the design will prove to be successful.

2

A change for the better – significance testing

2.1 Introduction

'Have we improved the process?' This is a frequently asked question. A change has been made to the process on the basis that it will improve it, but do subsequent results confirm this? It is perhaps usual for people to say, for example, that before the change the process gave a mean reading of 29.3, but now it is 29.6, therefore the change has worked. However, all processes are subject to noise (error, variation) and thus the change to 29.6 could be due to noise rather than the change. How are we to know? As well as quantifying the mean of a process we must also quantify the variability. Then we shall be in a position to know whether 29.6 could have been caused by noise or whether it must have been due to the process change because it is well beyond the value that could have been attributed to noise.

As well as looking at the above situation we shall take the opportunity in this chapter to introduce some of the building blocks of statistics used in analysing experiments – measure of average and variability, blob charts, histograms, normal distribution, confidence intervals and significance tests. However, these are very brief expositions of these topics and for a more in-depth treatment readers are referred to 'Statistical Methods in Practice' by Boddy & Smith.

2.2 Towards a darker stout

Stout is a drink that traditionally is dark – in fact the darker the better. Trentside Ales use adjunct AX751 to give a dark colour but a rival product BZ529 has been trialled on a pilot. The trials indicated that BZ529 gave a darker colour so Rob Whetham, the Development Manager of Trentside, has decided to put it into production in one of their four vats and then compare it with batches using AX751. The past 20 batches gave the results shown in Table 2.1 for darkness using AX751 as measured on a spectrometer:

Effective Experimentation: For Scientists and Technologists   Richard Boddy and Gordon Smith
© 2010 John Wiley & Sons, Ltd

Table 2.1   Darkness of 20 batches using AX751

Batch	1	2	3	4	5	6	7	8	9	10
Darkness	207	193	218	209	197	181	202	226	213	199
Batch	**11**	**12**	**13**	**14**	**15**	**16**	**17**	**18**	**19**	**20**
Darkness	206	215	204	205	189	201	220	211	186	194

Rob's first task is to look for trends since the presence of a trend would indicate the process was not stable and it would be difficult to judge whether the new additive had been effective. A runs chart for the data is shown in Figure 2.1 that indicates that the data has no trend. A better analysis would be using a cusum technique as outlined in *'Statistical Methods in Practice'*.

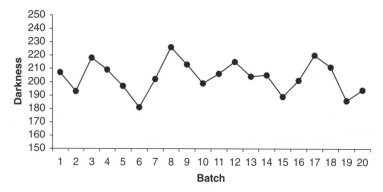

Figure 2.1   Runs chart.

Rob next looks at the distribution of data using a blob diagram as shown in Figure 2.2.

The distribution is typical of a process under control with values spread fairly symmetrically with more in the centre.

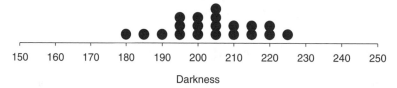

Figure 2.2   Blob diagram.

2.3   Summary statistics

The distribution of the data, as shown in a blob diagram or histogram, provides a valuable visual way of making judgements about the data. As well as the plots it is usually valuable to summarize data in terms of average and variability using the mean and standard deviation.

The mean is the sum of all observations divided by the number of observations.

The sum was 4076, there were 20 observations, so the mean in this case equals 203.8. It is usually referred to as the sample mean ($\bar{x}$).

The standard deviation can be defined as the average of the **deviations** from the mean. It is a strange sort of average but it does indicate the amount of variation. It is usually referred to as the sample standard deviation (s). Its value is 11.7. If we look at the 20 values we see that 14 are within one standard deviation of the mean, i.e. 192.1 to 215.5 and six are outside these limits which is to be expected when we use a standard deviation.

2.4   The normal distribution

Figure 2.3 shows the data presented in an alternative way to a blob diagram.

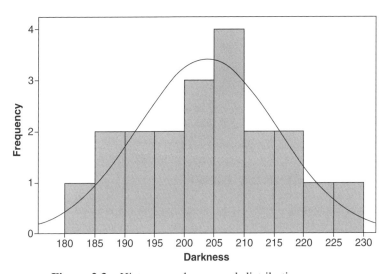

Figure 2.3   Histogram plus normal-distribution curve.

In a histogram, the height of each bar is proportional to the number of batches within the specified range. Superimposed onto the histogram is a normal-distribution curve. The normal distribution occurs when a process is subject to many additive errors, none of which are dominant in magnitude. It is found in abundance in processes where effort has been made to reduce the effect of any dominant source of error. It is also found in many natural processes.

The normal distribution is defined by two parameters, the population mean and population standard deviation. However, the population statistics are unknown and Rob has only sample data so it is these values – i.e. mean of 203.8 and a standard deviation of 11.7 that have been used to draw the normal distribution.

The normal distribution is often important in analyses of experiments since it provides a check on the validity of the conclusions. It is, however, very difficult to assess using histograms. A far better check is a probability plot shown in Figure 2.4. In this, the 20 observations have been ranked in order and then plotted against a nonlinear scale that represents the position of the observation if it was obtained from a perfect normal distribution.

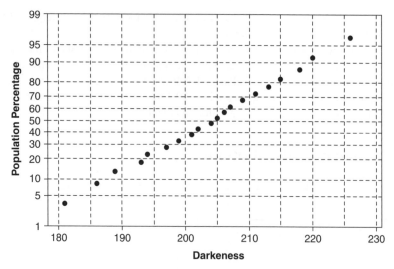

Figure 2.4   Normal probability plot.

The judgement we must now make, allowing for sampling or other errors, is: 'Is the shape of the plot sufficiently linear?' This takes judgement but Rob is assured that a straight line is a very good fit.

2.5   How accurate is my mean?

Rob has 20 batches with darkness values. Assuming these are representative samples from a stable process, how accurate is the mean? In order to calculate its accuracy we use a 95% confidence interval for the population mean (μ) that is given by the formula:

$$\overline{x} \pm \frac{ts}{\sqrt{n}}$$

where $\overline{x}$ is the sample mean

s is the sample standard deviation

n is the number of observations

and t is the coefficient obtained from Table A.2 using a 95% confidence level and $n-1$ degrees of freedom.

Degrees of freedom are the number of independent deviations from the mean used to calculate the standard deviation, one less than the number of observations. To illustrate this, if we have two results, say 240 and 260, the mean is 250 and both results must give the same deviation of 10. Thus, there is only one degree of freedom.

In our example:

$\overline{x} = 203.8$

$s = 11.7$

$n = 20$

$t = 2.09$ with 19 degrees of freedom at a 95% confidence level

$$95\% \text{ confidence interval} = 203.8 \pm \frac{2.09 \times 11.7}{\sqrt{20}}$$
$$= 198.3 \text{ to } 209.3.$$

This means we are 95% certain that this stable process will, in the long run, produce stout using AX751 with a mean darkness between 198.3 and 209.3.

Rob is pleased with the narrowness of the interval. This should allow him to make a reasonable judgement about whether the new additive is better.

2.6   Is the new additive an improvement?

Rob is now in a position to design and carry out an experiment. He decides to use six trials, not based on design considerations but on product considerations. If the adjunct BZ529 is detrimental to the product he would wish to abandon it fairly quickly.

The next six batches give darkness values as shown in Table 2.2.

Table 2.2   Darkness values of 6 batches from BZ529

Batch	21	22	23	24	25	26
Darkness	215	231	208	195	225	198

Rob is both encouraged and discouraged by the results. The result of 231 is higher than darkness readings obtained with AX751 but 195 and 198 are below the mean of the last 20 batches. On the other hand, the mean of BZ529 (212.0) is higher than obtained with AX751. A plot of the data for both adjuncts is shown in Figure 2.5.

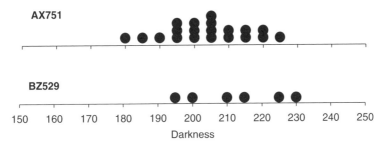

Figure 2.5   Blob diagram for the two adjuncts.

Because of variation any two means are always likely to be different. What we need to know is whether this difference is due to chance or is it due to the adjuncts? We can answer this by carrying out a two-sample t-test, but before doing so we need to combine the two standard deviations from the two adjuncts. The summary statistics for the two adjuncts are shown in Table 2.3.

Table 2.3 Summary statistics

	Mean	Standard deviation	Number of results	Degrees of freedom
Adjunct AX751	203.8	11.7	20	19
Adjunct BZ529	212.0	14.4	6	5

$$\text{Combined SD(s)} = \sqrt{\frac{(df_A \times SD_A^2) + (df_B \times SD_B^2)}{df_A + df_B}}$$

where SD_A, SD_B are the sample standard deviations for AX751 and BZ529 with df_A and df_B degrees of freedom, respectively

$$= \sqrt{\frac{(19 \times 11.7^2) + (5 \times 14.4^2)}{19 + 5}}$$

$$= 12.3 \text{ with } 19 + 5 = 24 \text{ degrees of freedom.}$$

We use the following procedure to carry out the significance test:

Null hypothesis: In the long run the two adjuncts will give the same mean, i.e. $\mu_A = \mu_B$

Alternative hypothesis: In the long run the two adjuncts will give different means, i.e. $\mu_A \neq \mu_B$

Test value: $= \dfrac{|\bar{x}_A - \bar{x}_B|}{s\sqrt{\dfrac{1}{n_A} + \dfrac{1}{n_B}}}$

where $\bar{x}_A$, $\bar{x}_B$ are the sample means for AX751 and BZ529, respectively; $|\bar{x}_A - \bar{x}_B|$ is the magnitude of the difference between $\bar{x}_A$ and $\bar{x}_B$; n_A, n_B the number of observations, and s the combined SD.

$$= \frac{|203.8 - 212.0|}{12.3\sqrt{\dfrac{1}{20} + \dfrac{1}{6}}}$$

$$= 1.43.$$

Table value: $t = 2.06$ from Table A.2 with $(n_A + n_B - 2) = 24$ degrees of freedom at a 5% significance level

Decision: If the test value is greater than the table value we reject the null hypothesis and accept the alternative.

However, in this case the test value is less than the table value. We cannot reject the null hypothesis.

Conclusion: There is insufficient evidence to conclude that there is a difference between the means for AX751 and BZ529.

Rob is disappointed. He has failed to establish that the new adjunct is significantly better but there are grounds for being optimistic – the sample mean is higher, the test value is approaching the table value and he only produced six batches with the new adjunct. How many batches should he have produced?

2.7   How many trials are needed for an experiment?

The number of trials required in a two-sample experiment is estimated by the formula:

$$n_A = n_B = 2\left(\frac{ts}{c}\right)^2$$

where n_A, n_B are the number of trials for each adjunct, c the difference that is required to be significant, s is a measure of the combined standard deviation, and t is a value from Table A.2 with the same number of degrees of freedom as the combined standard deviation.

In this example:

$s = 12.3$ and $t = 2.06$ from Table A.2 based on 24 degrees of freedom at a significance level of 5%

The difficult decision for Rob is the size of the difference (c) that needs to be found significant by the experiment. It is decided that improving the darkness on average by 8 will lead to a noticeable better stout.

$$n_A = n_B = 2\left(\frac{ts}{c}\right)^2 = 20.$$

We should note that this is a very useful formula, not only for a two-sample t-test but for factorial experiments given in later chapters. With two-level factorial designs this formula can be used to give an indication of the size of experiment required.

This is very convenient! Rob already has 20 batches with Adjunct AX751. He also has six batches with BZ529. He therefore produces 14 more batches that are added in Table 2.4 to those already produced.

Table 2.4   20 Batches with BZ529

Batch	21	22	23	24	25	26	27	28	29	30
Darkness	215	231	208	195	225	198	205	216	218	234
Batch	31	32	33	34	35	36	37	38	39	40
Darkness	229	209	211	220	217	193	205	222	199	216

Carrying out a significance test in the same manner as previously we use the same null and alternative hypotheses.

The summary statistics for BZ529 were: Mean $(\bar{x}) = 213.3$, SD $= 11.9$.

$$\text{Combined SD } (s) = \sqrt{\frac{(19 \times 11.7^2) + (19 \times 11.9^2)}{19 + 19}} = 11.8 \text{ with 38 degrees of freedom.}$$

$$\text{Test value} = \frac{|203.8 - 213.3|}{11.8\sqrt{\dfrac{1}{20} + \dfrac{1}{20}}} = 2.55$$

Table value $= 2.02$ from Table A.2 at a 5% significance level with 38 degrees of freedom.

Bob can conclude that BZ529 gives a significant improvement with an increase of 9.5 on the mean darkness.

2.8   Were the aims of the investigation achieved?

Yes.

Bob has shown, beyond reasonable doubt, that the new adjunct is significantly better. He also designed the investigation so the number of trials was chosen to find a difference that was materially important.

But...

We are dealing with a process that uses natural raw materials. How can we be sure the improvement was due to the adjunct and not due to a change in the materials?

Let us look at this question in more detail. There are three possible approaches:

To revert back to AX521 and determine whether the darkness returns to its original level. This is a well-based theoretical design but in practical terms it is a nonstarter. Why should Trentside Ales produce 20 batches of what they now believe will be an inferior product?

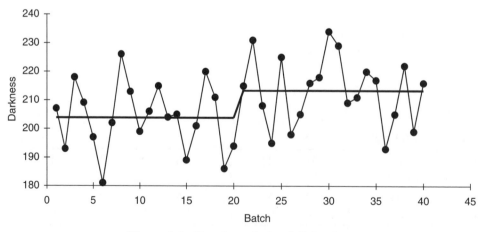

Figure 2.6   Trends analysis of 40 batches.

Change another vat from using AX751 to BZ529 and observe the difference in design. This is an excellent design if facilities are available but will prolong the changeover period to the superior adjunct.

Observe when the change in performance took place using cusum analysis. With this data it shows that the change took place at exactly batch 20 (in practice, we would expect it to be between 18 and 22) leading to further evidence that the change to BZ529 caused the improvement. The results of the cusum analysis is shown in Figure 2.6.

2.9   Problems

Jackson, the Senior Development Officer of Seltronics, had carried out an investigation into the performance of a prototype production line for 1000 µF capacitors. His investigation had found that the prototype production line was capable of making capacitors with a satisfactory average leakage current of below 100 µA. At this juncture, the Technical Manager of Seltronics draws Jackson's attention to a research article that indicates that a slight modification to the method of fabrication can achieve a worthwhile reduction in leakage current. He asks Jackson to investigate further.

Jackson decides to carry out the modification and produce a number of batches with the modified process. Jackson's first finding is that the modification is far from slight and will increase the running costs considerably. In Jackson's opinion it will be necessary to demonstrate a significant decrease in leakage current before the modification is installed on the full production line. Bearing this in mind he tests 8 capacitors, one from each batch, and obtains the results given in the table below and also shown as a blob diagram.

		Mean	Standard deviation
Standard process	113 65 43 32 78 59 142 43 95 71	74.1	34.32
Modified process	84 56 122 30 43 100 40 61	67.0	32.15

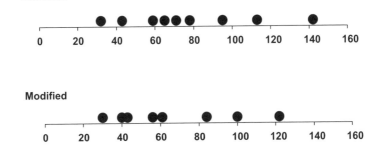

The table also includes results from the ten batches produced before the modification. For convenience we shall refer to these as being produced by the standard process.

Examining the blob diagrams it would appear that the standard process results do not come from a normal distribution. However, with so little data it is often difficult to make such a judgement using a blob diagram.

a) Use a normal probability plot on the standard process data and decide whether it fits a normal distribution.

b) Calculate 95% confidence intervals for the mean of each process.

c) Using a two-sample t-test decide whether the true means are significantly different.

3

Improving effectiveness using a paired design

3.1 Introduction

In Chapter 2 we were introduced to many aspects of design and also the use of significance testing to determine whether an improvement could have occurred by chance or really was due to a change in the process. In particular, we looked at a design for two independent samples and analysed the design using the two sample t-test for independent samples. In this chapter we look at a similar design – that is one with two samples – but where the observations are paired – that is linked by another variable. We also look at reducing variability and thereby making the design more efficient.

3.2 An example: who wears the trousers?

Crimpolock Ltd has long been known for stylish fabric for casual trousers. However, one frequent complaint has been that the Crimpolock fabric wears badly. To this end they instituted a research programme run by Caz Hewell that has shown that a new process will improve the wear. However, this improvement is purely based on laboratory tests that simulate real wear. These have been developed over a number of years and have been found to be useful at the process development stage. However, nothing replaces real wear and Caz now realizes that he must have two batches of fabric – one from the old process and one from the new – make these into trousers and find a group of volunteers who will trial trousers by wearing them. However, he knows the task is daunting since there are so many issues to consider:-

- How is the wear to be assessed?
- How often do you need to recall the trousers to assess the wear?

- How do you obtain the participants? How representative a sample should these be and what population should they represent?
- All wearers will wear trousers for different lengths of time and for different activities. This will lead to huge variability. How is this to be controlled?

Let us look at these issues in more detail.

3.3   How do we rate the wear?

This is a major problem since many features of wear can occur from broken threads, pulls and worn out fabric. If we are not careful the assessment will be very subjective. Caz therefore decides to find a set of fabrics (called standard fabrics) with a range of wear and rank these from 1 (least wear) to 10 (greatest wear). To rate each trouser the fabric will be compared against the standard fabrics and the assessor will give the rating that he or she believes is closest to the standard fabric. However, although using standard fabrics gives a scale it is not fully objective because of different wear features and thus Caz decides to use five assessors and average their ratings.

3.4   How often do you carry out an assessment?

Clearly carrying out an assessment is time consuming in terms of obtaining all the garments and then assessing them. If they are called in too quickly it will be a waste of time and effort since little wear will have occurred. If the period is too long the wear will be too great and the sensitivity of the test may have been lost. What length is required? To obtain a decent estimate past records should be looked at and the length of time (t) to obtain a wear rating of 7. The wear will be assessed at 3 periods – $t/3$, $2t/3$ and t. If sufficient wear has not been obtained at time t, another length of time will then be chosen.

3.5   Choosing the participants

Ideally the participants should be representative of all users by selecting members of the general public. However, Caz realises that this will require the involvement of a market research company and extensive resources at considerable cost. A more modest trial would be beneficial initially to indicate if the fabric is successful. They therefore decide to use volunteers from the workforce who would normally buy trousers made from the fabric.

3.6   Controlling the participants

This presents Caz with a real dilemma. How is he to control the participants? Does he stipulate how many hours per week they should wear the trousers and the acceptable activities? If he does this the participants may not be forthcoming. Another alternative is to ask the participants to record daily the length of time in use and the activities. However, this will put a great burden on the participants, the recording may not be done correctly and it may prove worthless for analysis purposes. Thus, attempts to control the participants are likely to be futile and, unless a full-scale market research trial is to be undertaken, this kind of trial should be avoided. Caz still needs to go ahead with the trial in some way but fortunately there is another alternative – a paired design.

3.7   The paired design

The trousers trial is ideal for a paired design. In this, each participant will be given a pair of trousers with one leg made from fabric from the new process and one from the old process. Just to eliminate bias Caz specifies that half the trousers should have the new process fabric on the right leg and half on the left leg. He obtains 20 volunteers who are regular wearers of trousers, gives each one a pair of the experimental trousers. Ten are chosen at random to receive the trousers with the old process on the right leg, the remainder receive the opposite. Caz keeps a record of which fabric is on which leg for each participant. He instructs them to wear the trousers frequently for six weeks before returning them for assessment. Notice the simplicity of instructions to the participants – no need to specify duration of wear or activity since obviously on any wearer both fabrics will get the same wear. Even better, by not specifying activities the fabrics will get a more representative sample of wear.

After six weeks the fabrics are assessed blind by the five assessors and the average scores are obtained. The fabrics are then decoded to give the values and the differences (Old – New) in Table 3.1.

Table 3.1   Wear ratings and differences

Participant	1	2	3	4	5	6	7	8	9	10
Old process	7.8	3.8	6.4	9.2	6.6	6.0	8.0	6.8	8.6	4.6
New process	7.2	4.2	5.0	8.2	6.0	5.4	6.4	6.4	7.2	3.4
Difference	0.6	–0.4	1.4	1.0	0.6	0.6	1.6	0.4	1.4	1.2
Participant	**11**	**12**	**13**	**14**	**15**	**16**	**17**	**18**	**19**	**20**
Old process	8.2	6.0	7.6	7.0	6.4	5.4	6.8	4.2	8.0	7.2
New process	7.0	6.0	6.6	5.8	6.6	6.4	7.0	3.8	7.8	7.6
Difference	1.2	0.0	1.0	1.2	–0.2	–1.0	–0.2	0.4	0.2	–0.4

The summary statistics are shown in Table 3.2.

Table 3.2   Summary statistics

	Mean	Standard deviation
Old process	6.73	1.45
New process	6.20	1.30
Difference	0.53	0.73

We can see that there is a high variability in wear in the old process that has a standard deviation of 1.45. Some of the variability is clearly caused by the assessment method but the main source must be the variability in use between the participants. The ratings for the new process have a similar level of variability to the old process. We notice that not all participants showed an improvement since some gave negative differences but the variability of the differences is much reduced with a standard deviation of only 0.73 despite it being obtained by subtracting two variable ratings. This is a clear sign that the pairing has been effective.

Is the mean difference of 0.53 significant? Let us perform the **paired t-test**.

Null hypothesis -The population means for the old process and the new process are equal, i.e. the mean of the population of differences is equal to zero ($\mu_d = 0$).

Alternative hypothesis -The mean of the population of differences is **not** equal to zero ($\mu_d \neq 0$)

Test value - $= \dfrac{|\bar{x}_d - \mu_d| \sqrt{n_d}}{s_d}$

where

$\bar{x}_d$ is the sample mean of differences $= 0.53$

μ_d is the mean difference according to the null hypothesis (0)

s_d is the sample standard deviation of the differences $= 0.73$

n_d is the number of differences $= 20$

$$= \frac{|0.53 - 0| \sqrt{20}}{0.73}$$
$$= 3.26.$$

Table values -   From Table A.2 with 19 degrees of freedom:
                      2.09 at the 5% significance level;
                      2.86 at the 1% significance level.

Decision -   We can reject the null hypothesis at the 1% significance level.

Conclusion -   The new process reduces wear.

The experimenter chose a significance level of 5% but as the test value of 3.26 was not only greater than the table value at 5% (2.09) but also greater than the 1% table value (2.86) he can be even more sure that there is a difference between the treatments, as his risk of making the wrong conclusion is less than 1%.

Thus, there is little doubt that the new process is better, but by how much? This can be obtained by weighing up the decrease in cost against the decrease in wear from the 95% confidence interval for the mean difference:-

$$\bar{x}_d \pm \frac{t s_d}{\sqrt{n_d}}$$

To summarize our data:

$\bar{x}_d = 0.53$ $n_d = 20$ $s_d = 0.73$ (19 degrees of freedom), $t = 2.09$ (5% level)

A 95% confidence interval for the population mean difference is:

$$0.53 \pm \frac{2.09 \times 0.73}{\sqrt{20}} = 0.53 \pm 0.34 = 0.19 \text{ to } 0.87.$$

Caz is pleased that the significance test has confirmed that the new process gives less wear but is disappointed that the confidence interval shows that the gain is small. In view of the number of participants with a fairly high rating for wear would a reduction of 0.19 (the lower end of the confidence interval) really lead to better customer satisfaction?

3.8   Was the experiment successful?

Yes.

Caz has carried out a highly successful experiment because he concentrated on a design that gave less variability in the results. He could have easily used a design for independent samples in which the variability would have swamped any improvement. He also ensured that the method of assessment was suitable for the purpose. The result was that he was able to show that the new process was better.

But...

Significance testing is necessary. It shows a change has not occurred by chance. However, it does not indicate that it is important. The magnitude of the improvement taken in conjunction with the confidence interval indicates that the improvement may be too small to improve customer satisfaction.

3.9   Problems

Jackson and Jackson deliver products to a large number of retailers. As part of their quality improvement programme they carry out customer satisfaction surveys. These are comprised of many questions referring to reliability, finish, returns, display units, delivery and telephone queries. Each answer is placed on a scale of 1 to 5. The total gives a final score that lies between 20 and 100. The scores for the last two quarters are given below with the same retailers being used in each quarter.

Results from a paired sample survey

Customer	A	B	C	D	E	F	G	H	Mean	SD
1st quarter	68	82	87	72	80	83	94	66	79.0	9.67
2nd quarter	63	75	85	73	71	78	88	59	74.0	9.93

1. Carry out a paired t-test.

2. What sample size is required to detect a difference of 3.0?

3. Using the above results and the methods outlined in Chapter 2, how many samples would be required to detect a mean difference of 3.0 if the results had been independently obtained in each quarter, i.e. not from the same retailers.

4

A simple but effective design for two variables

4.1 Introduction

In the last two chapters we have looked at designs involving one variable or factor such as two different adjuncts or processes. We concentrated on making sure the design was valid and the sample size was sufficient to draw a meaningful conclusion. We now move on to consider designs with more than one variable. These are referred to as factorial designs. Factorial experiments are relatively simple to understand, and yet their simplicity should not cause us to overlook their powerfulness. They allow us to unravel complex relationships between a response variable and several independent variables (or factors) without needing to do complex analysis, because the effects of all the variables (and their interactions) can be measured independently of each other. Factorial designs, as we shall see, are also extremely efficient designs.

4.2 An investigation

Before considering a simple factorial experiment, we will review an experiment that was carried out by Hamish Ons, a research scientist employed by Grant Oils. The current project on which Hamish is working seeks to improve the fuel economy (miles per gallon – mpg) of Grant's best-selling Slick Oil. There are many components to the formulation of an oil, but Hamish has identified two that he believes may have an effect on the fuel economy – **friction modifier** and **type of base oil**. We shall refer to these as independent variables. He will plan an experiment in which he varies these components and hopes to determine which formulation gives the best fuel economy. Each trial will involve a test car, with the same driver, being driven for ten circuits of the company's test track using the same cycle of speeds and the mean mpg determined.

Effective Experimentation: For Scientists and Technologists   Richard Boddy and Gordon Smith
© 2010 John Wiley & Sons, Ltd

Hamish decided that the friction modifier needs to be at a treat rate between 0.3% and 0.6% and he would investigate two base oils – mineral and synthetic. The present oil uses a mineral-base oil and friction modifier at 0.3%.

4.3   Limitations of a one-variable-at-a-time experiment

Hamish proceeded as follows:

Stage 1: To estimate the effects of changing base oils, Hamish produced two blends at a friction modifier treat rate of 0.3%. Four tests were carried out on each blend. The results are shown in Table 4.1.

Table 4.1 Mean fuel economy by base oil type. (With Friction Modifier 0.3%)

	Mineral	Synthetic
	34.5	35.7
	35.1	35.2
	35.3	36.1
	34.9	35.8
Mean	34.95	35.70
Standard deviation	0.342	0.374

Hamish was very pleased with the outcome of this part of the experiment. Applying a two-sample 't'-test to the above data shows that the synthetic base oil gave a significantly higher fuel economy than the mineral-base oil.

(Test value $= 2.96$; table value at the 5% significance level from Table A.2 with 6 degrees of freedom $= 2.45$.)

Stage 2: To estimate the effect of changing the friction modifier, Hamish decided to carry out 4 more tests using the synthetic base oil with friction modifier treat rate of 0.6%. The fuel economy of these tests, together with the data obtained previously for the synthetic-base oil, are shown in Table 4.2.

Table 4.2 Mean fuel economy by friction modifier (with synthetic base oil)

	Friction modifier	
	0.3%	0.6%
	35.7	36.6
	35.2	35.8
	36.1	36.1
	35.8	36.5
Mean	35.70	36.25
Standard deviation	0.374	0.370

Hamish was less satisfied with the outcome of this part of the experiment, because though it appears that the higher treat rate for friction modifier produces better fuel economy, there is not sufficient evidence from the experiment to be certain that this is the case. (A two-sample 't'-test gives: Test value $= 2.09$; 5% table value from Table A.2 with 6 degrees of freedom $= 2.45$.)

At this point it is appropriate to review the results of this experiment. From Table 4.1 we can estimate the **effect** of base oil type by:

$$\text{Effect of base oil} = \text{Mean fuel economy using synthetic}$$
$$-\text{Mean fuel economy using mineral}$$
$$= 35.70 - 34.95$$
$$= 0.75 \text{ mpg}$$

The base-oil effect suggests that using synthetic rather than mineral increases the fuel economy, on average, by 0.75 mpg.

We may also compute a confidence interval for this effect estimate (a confidence interval for the difference between two population means):

$$95\% \text{ confidence interval for the base-oil effect} = (\bar{x}_B - \bar{x}_A) \pm ts\sqrt{\frac{1}{n_A} + \frac{1}{n_B}}$$

(where $s =$ combined SD estimate $= 0.358$)

$$= (35.70 - 34.95) \pm \left(2.45 \times 0.358 \times \sqrt{\frac{1}{4} + \frac{1}{4}}\right) = 0.75 \pm 0.62$$

(The 0.62 is sometimes referred to as the least significant difference.)

We are 95% confident that the true effect of using synthetic rather than mineral is to increase the fuel economy by between 0.13 mpg and 1.37 mpg.

Using the results from Table 4.2 we can estimate the **effect** of the friction modifier by:

$$\text{Effect of friction modifier} = \text{Mean fuel economy at high \% friction modifier}$$
$$-\text{Mean fuel economy at low \% friction modifier}$$
$$= 36.25 - 35.70$$
$$= 0.55 \text{ mpg}$$

We compute the 95% confidence interval for this effect as before, to obtain

$$95\% \text{ confidence interval for the friction-modifier effect} = 0.55 \pm 0.64 \text{ mpg}$$

Hence, we are 95% confident that the true effect of increasing the % friction modifier by 0.3% is to **increase** the fuel economy on average by between -0.09 mpg and $+1.19$ mpg. Note that the confidence intervals are in total agreement with the results of the two-sample 't'-tests: the base-oil effect is definitely **not** zero but is positive, while the friction-modifier effect **could be** zero.

What has Hamish achieved from this series of experiments?

He has established that by changing from the mineral to the synthetic-base oil he can significantly improve the fuel economy. However, this was only with 0.3% friction modifier.

Increasing the friction modifier treat rate from 0.3% to 0.6% made no significant difference to fuel economy. This was by continuing with the mineral-base oil.

Hamish was somewhat perplexed as to what to do next. He suspected that if he were to carry out four more tests with synthetic and 0.6% friction modifier he may generate sufficient evidence to show that the friction-modifier effect was also significant. However, he was not certain whether this would be worthwhile given the data already available. Hamish was also under pressure from the Marketing Manager to complete his experiment quickly, because the Marketing Manager had to launch the best formulation as a new product. Before proceeding, however, he plotted the results on a graph as shown in Figure 4.1.

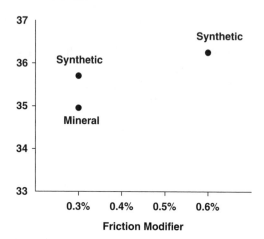

Figure 4.1   Data from stages 1 and 2 of the experiment.

He then thought it would be useful to add to the graph his estimate of the relationship between fuel economy and friction modifier. He added the following lines to the graph (Figure 4.2):

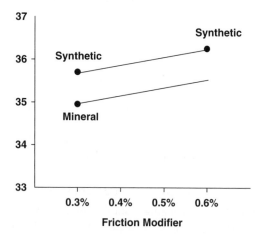

Figure 4.2   Hamish's estimated relationship between fuel economy, friction modifier and base oil type.

In one respect the lines on the graph are helpful. One line allows the eye to follow the change between the two levels of friction modifier, but the graph could be misleading because the other line assumes that there is no interaction between friction modifier and base oil type, that is, the effect of friction modifier will be the same irrespective of the choice of base oil. If this assumption is not valid and there is an interaction between friction modifier and base oil type, the graphs could be like either of those in Figure 4.3.

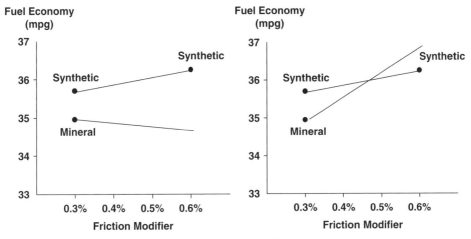

Figure 4.3   Possible relationships if an interaction exists.

An **interaction** occurs when the effect on the response of one independent variable is dependent upon the level of another independent variable.

In Figure 4.2, Hamish has shown a situation with **no interaction**. He is assuming that the effect of friction modifier on fuel economy is the same for both base oils.

The graphs in Figure 4.3 show two situations in which an interaction exists. In both cases, the effect of friction modifier on fuel economy depends upon the base oil type.

An interaction is a **state of nature.** If interactions exist it is important that they are identified. Unfortunately, the one-variable-at-a-time approach does not usually lead to the discovery of interactions. However, factorial designs enable the experimenter to estimate both the effects of the main factors and the interactions between these factors.

Hamish realizes that he cannot really tell anything about the effect of base and friction modifier on the fuel economy unless he has data from mineral oil at 0.6% friction modifier.

He gives more thought to what is needed before starting the next stage. He realizes that he should carry out trials at all combinations of base oil and friction modifier, using a factorial experiment.

4.4   A factorial experiment

As Hamish cannot be certain whether there will be an interaction between base oil type and friction modifier, he undertakes another experiment. The design is shown in Table 4.3.

We can easily see that this experiment includes the three trial conditions that Hamish has used previously, but also includes a fourth combination that completes the overall balance of the experiment. There are **two levels** of both factors, and **all** of the possible combinations

Table 4.3 A simple factorial experiment

Base oil	Friction modifier	
	0.3%	0.6%
Mineral	X X	X X
Synthetic	X X	X X

of the levels of both factors are used. With **two levels** of each of two factors we have a 2^2 **experiment**. A 2^2 experiment uses 4 treatment combinations. Hence, a simple 2^2 experiment can be undertaken using only 4 trials, but Hamish decides to replicate the experiment by using each treatment combination more than once.

(Note the notation: an experiment using 3 levels for each of 4 factors is a 3^4 factorial experiment. Such an experiment uses $3^4 = 81$ treatment combinations.)

Hamish undertakes the new experiment. However, Hamish is not sure that 8 trial batches (i.e. 2 under each set of treatment conditions) will be sufficient, but he is under pressure and knows that no more time or resources are available so he goes ahead.

When the experiment is complete, the results are as shown in Table 4.4.

Table 4.4 Results of 2^2 factorial experiment (mean fuel economy)

Base oil	Friction modifier	
	0.3%	0.6%
Mineral	34.8	36.7
	35.1	37.2
Synthetic	35.5	36.4
	35.9	36.1

To calculate the main effects (and the interaction effect) it is useful to summarize the above data by the means for each treatment combination and the means for the rows and columns, as shown in Table 4.5.

Table 4.5 Row, column and cell means

Base oil	Friction modifier		
	0.3%	0.6%	Mean
Mineral	34.95	36.95	35.95
Synthetic	35.70	36.25	35.97
Mean	35.32	36.60	35.96

Comparing the **row means:**

$$\text{Base-oil effect} = \text{Mean fuel economy using synthetic}$$
$$-\text{Mean fuel economy using mineral}$$
$$= 35.97 - 35.95$$
$$= 0.02 \text{ mpg.}$$

Comparing the **column means**:

$$\text{Friction-modifier effect} = \text{Mean fuel economy at high \% friction modifier}$$
$$-\text{Mean fuel economy at low \% friction modifier}$$
$$= 36.60 - 35.32$$
$$= 1.28 \text{ mpg.}$$

In general we refer to the effect of a variable as its **main effect** to distinguish it from interaction effects that we will also estimate. We always calculate the **main effect** as follows:

Main effect of a variable = Mean response at the high level of the variable
−Mean response at the low level of the variable

For these variables, which is the high level and which is the low level?

For % friction modifier, it is easy to tell. 0.6% is higher than 0.3% so it is the high level.

For base oil, we have a choice of two oils in the experiment. As we are interested in the change in fuel economy, it is usual to calculate the difference as we go from present conditions to new conditions, so we would treat the present condition, mineral oil, as the low level and the new condition, synthetic, as the high level. Often the choice is arbitrary, but it is then important that the choice is maintained throughout the analysis.

Comparing the cell means we notice an inconsistency:

(a) with mineral, an increase in friction modifier of 0.3% (from 0.3% to 0.6%) causes an increase in fuel economy of 2.0 mpg;

(b) with synthetic, an increase in friction modifier of 0.3% causes an increase in fuel economy of only 0.55 mpg.

This 'inconsistency' indicates that there is an **interaction** between base oil type and friction modifier. The effect of friction modifier on fuel economy depends on the type of base oil.

We can of course also see the interaction if we compare the base-oil effects at different friction modifier levels:

(a) with a friction modifier of 0.3%, synthetic gives higher fuel economy;

(b) with a friction modifier of 0.6%, mineral gives higher fuel economy.

In general, an interaction effect is defined as follows:

Interaction = ½(Effect of Factor A at high level of B
− Effect of Factor A at low level of B)

The interaction of friction modifier and base oil type is

$$\tfrac{1}{2}(\text{Effect of friction modifier using synthetic}$$
$$-\text{Effect of friction modifier using mineral})$$
$$= \tfrac{1}{2}(0.55 - 2.00)$$
$$= -0.72$$

However, it is more convenient to calculate the interaction from –

$$Interaction = Difference\ between\ diagonal\ means$$

which always gives the same result.

The direction of the difference is important:

$$Interaction = Mean\ on\ leading\ diagonal - Mean\ on\ secondary\ diagonal$$

The diagonal means are indicated in Table 4.6. The leading diagonal goes from top left to bottom right.

Table 4.6 Cell and diagonal means

Base oil	Friction modifier		
	0.3%	0.6%	
			36.32 (Sec.)
Mineral	34.95	36.95	
Synthetic	35.70	36.25	
			35.60 (Leading)

Thus the interaction is calculated by

$$Interaction = Mean\ on\ leading\ diagonal - Mean\ on\ secondary\ diagonal$$
$$= 35.60 - 36.32$$
$$= -0.72.$$

We can now bring together the results of our calculations that have given us three estimates:

$$Base\text{-}oil\ effect = \quad 0.02$$
$$Friction\text{-}modifier\ effect = \quad 1.28$$
$$Base\ oil \times friction\ modifier\ interaction\ effect = -0.72$$

We can interpret these estimates as follows:

(a) The effect of changing the base oil from mineral to synthetic is to increase the fuel economy by 0.02 mpg, on average.

(b) The effect of increasing friction modifier from 0.3% to 0.6% is to increase the fuel economy by 1.28 mpg, on average.

(c) There is an interaction between base oil type and friction modifier. The existence of this interaction nullifies the value of statements (a) and (b) above. These statements are not incorrect, but they are worthless and potentially misleading. **Whenever an interaction exists it is unwise to speak of either of the two main effects in isolation.** The existence of the interaction tells us that the effect of friction modifier depends on the base oil type. The relationship between the three variables – base oil type, friction modifier and fuel economy – is best described with reference to Table 4.5.

4.5   Confidence intervals for effect estimates

We have now estimated the main effects and the interaction effect, but, as yet, we have no idea about the **precision** of these estimates. It would be advisable to calculate a confidence interval for each effect.

In order to proceed we need an estimate of the **residual standard deviation** (in this case the batch-to-batch standard deviation of mean fuel economy). There are several ways in which this estimate could be obtained:

1. By calculating the standard deviation of the mean fuel economy of a large number of tests on a reference oil.

2. By using results from a previous experiment in which several tests under **fixed** conditions were undertaken (for example, from Hamish's original experiment).

3. By carrying out several tests on a reference oil in addition to a factorial experiment.

4. By **replicating** the trials in a factorial experiment, i.e. using each treatment combination **twice** (or three times). The trials should be undertaken in a predetermined and systematic order that causes no bias in the results.

In this case, Hamish has replicated the factorial experiment since he carried out two tests on each blend. To obtain an estimate of the residual standard deviation we first compute the cell standard deviations, i.e. the standard deviations of each pair of results produced using the same conditions. These are shown in Table 4.7.

Table 4.7   Cell standard deviations

Base oil	Friction modifier	
	0.3%	0.6%
Mineral	0.212	0.354
Synthetic	0.283	0.212

If we assume that the batch-to-batch variability of mean fuel economy does not depend on the experimental conditions then each cell standard deviation is an estimate of the batch-to-batch or residual standard deviation, based on 1 degree of freedom.

To obtain a better estimate, we may combine the four separate estimates using

Combined standard deviation

$$= \sqrt{\frac{\sum(df \times s^2)}{\sum(df)}}$$

$$= \sqrt{\frac{(1 \times 0.212^2) + (1 \times 0.283^2) + (1 \times 0.354^2) + (1 \times 0.212^2)}{1 + 1 + 1 + 1}}$$

$= 0.272$ based on 4 degrees of freedom

Our best estimate of the residual standard deviation (RSD) is thus 0.272. If we were to carry out a series of tests on a reference oil then we would expect the standard deviation of fuel economy to be approximately 0.272 mpg.

This is not a very good estimate of batch-to-batch variability, of course, as it is based on only four degrees of freedom.

To calculate the confidence intervals for the main effects and interaction use:

Confidence interval for any effect in a two-level factorial experiment:

$$\text{Effect} \pm \frac{2ts}{\sqrt{n}}$$

where n = total number of trials in the experiment

s = residual standard deviation

t = value from the 't'-distribution with the same degrees of freedom as s

In this experiment, Hamish has undertaken 8 trials and the RSD is 0.272 (with 4 degrees of freedom). The t value for a 95% confidence interval is thus 2.78 (with 4 degrees of freedom), and the confidence interval width will be

$$\pm \frac{2ts}{\sqrt{n}} = \frac{2 \times 2.78 \times 0.272}{\sqrt{8}}$$
$$= \pm 0.53$$

The full analysis may then be summarized as follows:

$$\text{Effect of base oil} = \quad 0.02 \pm 0.53$$
$$\text{Effect of friction modifier} = \quad 1.28 \pm 0.53$$
$$\text{Interaction effect} = -0.72 \pm 0.53$$

We can therefore conclude that the interaction and the friction-modifier effect are significant, whereas the effect of base oil type is not. As previously stated, where an interaction exists it is unwise to speak of the main effects in isolation. So when an interaction is significant, it is usually irrelevant to consider the significance of its main effects.

The confidence interval brings a valuable touch of reality to the quantification of the estimates. The friction-modifier effect could be as small as 0.75 mpg or as large as 1.81 mpg. It is interesting to note, however, that the confidence intervals (i.e. the precision) obtained from this factorial experiment with eight trials are very similar to those obtained from Hamish's original experiment.

Let us summarize the two experiments:

Experiment 1: one-variable-at-a-time
- 12 tests undertaken
- 2 main effects estimated
- precision of each estimate based on 8 trials (6 degrees of freedom)
- no estimate of interaction (and no indication whether there is an interaction).

Experiment 2: factorial design
- 8 tests undertaken
- 2 main effects and the interaction estimated
- precision of all estimates based on all 8 trials (4 degrees of freedom)

This summary reveals the efficiency of factorial designs. From the second experiment, Hamish Ons can obtain estimates of **more** effects (including the all-important interaction) with almost the **same level of precision**, but with **fewer** trials.

The 'balance' within the factorial design means that the result of each test in the experiment contributes to the estimate of each main effect and interaction. All 8 tests are used to estimate the base oil effect, but they are also all used to estimate the friction modifier effect, and to estimate the interaction effect.

On the other hand, the effects of base oil type and friction modifier are estimated completely independently of each other. The estimate of the friction-modifier effect is based on four tests at 0.3% and four tests at 0.6%. However, each set of four tests included two using mineral and two using synthetic. Hence, the base oil effect is averaged out in each set of four tests, and the estimate of the friction-modifier effect is independent of base oil type.

In summary, factorial designs are simple in concept, efficient in operation and provide independent estimates of all main effects and interactions.

4.6   What conditions should be recommended?

The recommended conditions should be based on the predicted values and their associated confidence intervals for a series of conditions. The computation of such values is given in Chapter 5. However, for the moment we can use the values obtained in Table 4.6 that represent the predicted values at the four conditions used in the experiment. It would be unwise to extrapolate outside the range of conditions or to interpolate within them.

Table 4.6 showed that whereas synthetic oil is considerable better with 0.3% friction modifier, mineral oil is marginally better at 0.6%.

This conclusion astounds Hamish. Synthetic oil is much more expensive than mineral, but this experiment shows that a high level friction modifier with the mineral-base oil is just as good if not better than the synthetic.

4.7   Were the aims of the investigation achieved?

Yes.

A simple factorial experiment with only four different formulations was able to estimate the effects of the two factors and their interaction.

Since the experiment was balanced, the estimates of the effects were obtained independently of each other and could be obtained using a simple formula.

At each set of conditions, two trials were run, so the residual standard deviation could be obtained, measuring the batch-to-batch variation.

The residual standard deviation allowed a confidence interval to be determined for each effect and the significance evaluated.

The factorial experiment achieved more that the series of experiments changing one variable at a time – all the estimated effects and the same level of precision with fewer trials.

From the factorial experiment, Hamish was able to establish that increasing the treat rate of the friction modifier improved the fuel economy, but the amount of improvement depended on whether the base oil was mineral or synthetic, that is, there was an interaction between friction modifier and base oil. He was able to conclude that the best formulation was mineral oil and 0.6% friction modifier.

But ...

This design is for only two factors, each of which has only two levels. It served its purpose for what was required, but most experiments will require a larger design. These are considered later in the book.

Did Hamish think of practical considerations in his experiments?
What order did he conduct the trials in?

Many industrial processes suffer from trends or step changes. If he had run them in the order implied by the table and there had been underlying changes, then any observed effect could have been spurious.

This experiment involved actual runs of a car on a circuit, so he should have ensured that any other effects (e.g. weather, state of the ground) were balanced and that the engine was completely flushed out between trials.

Hamish has found an astounding conclusion. Perhaps he would be wise to repeat the experiment, not necessarily with replicates, on another vehicle to confirm the conclusion.

No other response has been examined. This is not sensible. For example, a friction modifier at 0.6% with a mineral oil might cause excessive wear.

4.8   Problems

1. An experiment was carried out to investigate the effect of two variables - resin and cement concentrations - on the crushing strength of a grout mix. Altogether eight batches of grout mix were made and the following crushing strengths obtained:

Resin	Cement	
	10%	12%
1.0%	524	548
	536	552
1.5%	546	530
	538	540

   (a) How many independent variables are included in the experiment?
   (b) How many levels has each independent variable?
   (c) How many times is each cell replicated?
   (d) On the basis of your answers to the previous sections what kind of experiment has been used?
   (e) Calculate the mean and standard deviation for each cell.
   (f) Calculate the main effects and the interaction effect.
   (g) Calculate the residual standard deviation.
   (h) Calculate the 95% confidence interval for the effects and hence decide which effects are significant.
   (i) What conditions would you recommend for future production?

2. Dr. Scratchplan, a research scientist at a chemical plant, was conducting a series of experiments to determine conditions which would improve the plasticity of a polymer. He

was unwilling to explore the complete range of the variables until he had made a preliminary investigation over a narrower range around the present operating conditions.

He decided to use initial values as follows:

Pressure (P)	1.3 and 1.6 atm
Temperature (T)	271 °C and 279 °C
Agitation speed (S)	50 and 60 r.p.m.

In his first preliminary investigation he decided to do a 2^2 factorial experiment varying only two of the possible factors, pressure and temperature. To obtain an estimate of the error he decided to duplicate each of the four trials.

His results were as follows:

P	T	Plasticity	Plasticity	
			$\bar{x}$	s
1.3	271	213		
1.3	271	225	219	8.5
1.3	279	257		
1.3	279	245	251	8.5
1.6	271	258		
1.6	271	266	262	5.7
1.6	279	281		
1.6	279	289	285	5.7

(a) Analyse the results of the experiment. Which of the variables (main effects and interaction) are significant?
(b) After analysing the results Dr Scratchplan discovered that because no information was given to the plant supervisor about agitation speeds, he varied them using his own initiative.

The pattern of speeds he used with the combinations of temperature and pressure was:

	P	
T	1.3	1.6
271	50 r.p.m.	60 r.p.m.
279	60 r.p.m.	50 r.p.m.

How does this throw doubt on your original conclusions?

5

Investigating 3 and 4 variables in an experiment

5.1   Introduction

In Chapter 4 we saw how to design and analyse a two-variable factorial experiment with each of the variables at two levels. The example illustrated one of the main advantages of factorial experiments – their ability to diagnose and analyse interactions. However, another advantage was not well illustrated by the example. This is the efficiency of the design in obtaining accurate estimates of an effect. For example, the confidence interval for an effect for a one-variable-at-a-time approach involving 16 trials will be more or less the same as for a 2^4 experiment. With the former we can only investigate one variable, while with the latter we are able to investigate four variables and also a number of interactions. In this chapter we look at both a 2^3 and 2^4 experiment that will amply demonstrate the efficiency of these designs.

5.2   An experiment with three variables

Seltronics make many millions of transistors per year. Recently, the management of Seltronics have decided to tackle an endemic problem – that of the abysmal defective rates, currently running at 45%, by promoting Jones to the post of development engineer. Defectives are found by circuit failures at the final inspection in which every transistor is automatically tested. By examining past records for each batch Jones has established beyond doubt that the main factor affecting defectives is the thickness of the tin surface coating, which is applied by electroplating. Jones believes that the thickness should be between 600 and 900 μm and any thicknesses outside this range will lead to transistor failure.

The actual coating operation involves passing a jig containing 1000 transistors through an electroplating bath. Jones knows how to change the bath conditions so as to alter the mean coating thickness. He also knows that there is little difficulty in obtaining a mean of about 750 μm, the middle of the specification. However, the variability of coating thickness in each

Effective Experimentation: For Scientists and Technologists   Richard Boddy and Gordon Smith
© 2010 John Wiley & Sons, Ltd

jig is far too great, resulting in between 20% and 50% of the transistors per batch having coating thicknesses either below 600 μm or above 900 μm. For convenience we shall refer to this statistic as per cent defectives.

Now, the bad thing from Seltronics' point of view is the large % defective; the good thing from Jones's point of view is the variability in % defective from batch to batch. Clearly, if some batches have 50% defectives while others only have 20% it may be possible to find operating conditions that are likely to give around 20%, and perhaps by moving to appropriate conditions he can lower this value even further. Jones decides to learn more about the process and in doing so is led to believe that three variables – current, bath temperature and tin concentration – are important.

He will design an experiment to enable him to determine the effects of current, bath temperature and tin concentration, and their interactions, on per cent defectives and determine those conditions that will minimize the per cent defectives.

After much thought he decides to carry out a 2^3 factorial experiment with the following conditions:

Variable	Low level	High level
Current in Amps (A)	40 A	50 A
Bath temperature (B)	25 °C	30 °C
Tin concentration (C)	28 g/kg	32 g/kg

The results of the experiment are shown in Table 5.1.

Table 5.1   Three-way table of per cent defectives

A	Low C		High C	
	Low B	High B	Low B	High B
Low	46.3	44.1	21.4	22.7
High	36.2	36.4	40.8	39.3

From Table 5.1 we could calculate estimates of the three main effects using the formula that we used with a 2^2 design in Chapter 4:

Main effect of a variable = Mean response at the high level of the variable
−Mean response at the low level of the variable.

It would not, however, be obvious how we could calculate estimates of the interaction effects from this table, so we will first break down the three-way table into a two-way table, relating to B and C only. This is achieved by averaging the response values in pairs to give the mean responses at each combination of level of B and C as shown in Table 5.2.

From the two-way table we can calculate

$$\text{Effect of } B = 35.625 - 36.175 = -0.55$$
$$\text{Effect of } C = 31.05 - 40.75\quad = -9.70$$
$$\text{Interaction } BC = 36.125 - 35.675 = \quad 0.45$$

Table 5.2   Mean responses for B and C combinations

Bath temperature (B)	Tin concentration (C)			
	Low	High	Mean	
				35.675
Low	41.25	31.1	36.175	
High	40.25	31.0	35.625	
Mean	40.75	31.05		
				36.125

Now, to obtain all the other two-variable interactions we have to compute two more tables. After that we will only have the three-variable interaction (ABC) to compute. This can be defined as half of 'interaction BC at high level of A' minus 'interaction BC at lower level of A'. Clearly, this will be somewhat tedious to calculate and we must now realize that using two-way tables is not suitable with more than two variables.

We shall demonstrate a simpler, but more abstract method, called the design matrix. It is based upon the fact that every effect (main effect or interaction) may be calculated as 'the mean of all results at the high level of the effect' minus 'the mean of all results at the low level'. So, for every effect, half the results have a plus sign associated with them and the other half have a minus sign. The 'design matrix' method gives a simple way of deciding how these plus and minus signs are allocated.

5.3   The design matrix method

To apply the design matrix method, we first need to tabulate the design and the results from Table 5.1 in a different manner as shown in Table 5.3. The results are presented in a column, with the levels of the factors in the corresponding rows to the results.

Table 5.3   Design and responses for a 2^3 experiment

Current (A)	Bath temperature (B)	Tin concentration (C)	% defectives
40	25	28	46.3
50	25	28	36.2
40	30	28	44.1
50	30	28	36.4
40	25	32	21.4
50	25	32	40.8
40	30	32	22.7
50	30	32	39.3

In the order in which the results are tabulated, Table 5.3 shows a simple pattern for the independent variables. A alternates with low and high values, B alternates with two low and two high values, while C alternates with four low and four high values. Clearly this pattern will be the same for all 2^3 experiments and we can generate a general pattern using two symbols: '−' to represent the low level and '+' for the high level. This is shown in Table 5.4. When the trials are tabulated in this order, it is called the 'standard order' of the design matrix.

Table 5.4 Design matrix for a 2^3 experiment

A	B	C	% defectives
−	−	−	46.3
+	−	−	36.2
−	+	−	44.1
+	+	−	36.4
−	−	+	21.4
+	−	+	40.8
−	+	+	22.7
+	+	+	39.3

The effect of C is the mean response for all the pluses less the mean response for all the minuses, i.e.:

$$\left(\frac{21.4+40.8+22.7+39.3}{4}\right)-\left(\frac{46.3+36.2+44.1+36.4}{4}\right)$$

$$= 31.05 - 40.75$$
$$= -9.70$$

Similarly, the effects of A and B are obtained in the same way matching up the data with the '+' and '−' signs.

So far, this is nothing new apart from the notation, but the advantage of the design matrix method is that we can obtain the interactions in the same way. First we must obtain the analysis matrix that contains additional columns representing the interactions.

We see that the signs in each row of column AB are obtained by multiplying the corresponding row signs in columns A and B.

If we look at the levels in a the two-way table we can see that it is reasonable to obtain the signs for the interaction by multiplying the signs of the main factors:

Signs for AB interaction

A		B	
		−	+
−	+		−
+		−	+

The 'plus' signs are in the cells on the leading diagonal, and the 'minus' signs are on the secondary diagonal.

Likewise, the columns of the other interactions are obtained by multiplication of the component variables. We can now use Table 5.5 to calculate any of the interactions.

Table 5.5 Analysis matrix for a 2^3 design

A	B	AB	C	AC	BC	ABC	% defectives
−	−	+	−	+	+	−	46.3
+	−	−	−	−	+	+	36.2
−	+	−	−	+	−	+	44.1
+	+	+	−	−	−	−	36.4
−	−	+	+	−	−	+	21.4
+	−	−	+	+	−	−	40.8
−	+	−	+	−	+	−	22.7
+	+	+	+	+	+	+	39.3

For example interaction AC is given by

$$\left(\frac{46.3 + 44.1 + 40.8 + 39.3}{4}\right) - \left(\frac{36.2 + 36.4 + 21.4 + 22.7}{4}\right)$$
$$= 13.45$$

The main effects and interactions can be computed in a similar manner to give the values in Table 5.6.

Table 5.6 Estimates of main effects and interactions

Effect	Estimate	95% confidence interval
A	4.55	±2.02
B	−0.55	±2.02
AB	−0.10	±2.02
C	−9.70	±2.02
AC	13.45	±2.02
BC	0.45	±2.02
ABC	−1.30	±2.02

Overall mean = 35.9

Included in Table 5.6 is the 95% confidence interval. This has **not** been computed from the responses and we shall return to its calculation later. First, let us examine the results of the experiment. Judged by the 95% confidence interval there is no doubt about the effects of A and C and the interaction AC, since they cannot be equal to zero. All the other effects are not significant and we could well decide to ignore them.

5.4 Computation of predicted values

We have rightly decided to confine our interest to main effects A and C and the interaction AC. Thus, there are only 4 combinations of A and C for which we require to obtain predicted values. These are calculated in the following manner:

For example, with A at a high level and C at a low level the matrix levels are $A(+)$ and $C(-)$.

With those levels the interaction's level is $AC(-)$.

Since each effect is calculated by the difference between the mean response at high and low levels, to predict the response at each set of conditions we start with the overall mean response and half of the effect in the appropriate direction thus:

$$\text{Predicted value} = \text{Overall mean} + \tfrac{1}{2}(\text{Effect of } A \text{ -Effect of } C - \text{Interaction } AC)$$

$$= 35.90 + \tfrac{1}{2}(4.55 - (-9.70) - 13.45)$$

$$= 36.30$$

Similarly we can compute values for the other combinations of A and C to give predicted values, confidence intervals, and residuals as shown in Table 5.7.

Table 5.7 Predicted values and residuals

A	C	Predicted response ± confidence interval	Actual responses	Residuals
Low	Low	45.20 ± 2.0	46.3, 44.1	$+1.10, \; -1.10$
High	Low	36.30 ± 2.0	36.2, 36.4	$-0.10, \; +0.10$
Low	High	22.05 ± 2.0	21.4, 22.7	$-0.65, \; +0.65$
High	High	40.05 ± 2.0	40.8, 39.3	$+0.75, \; -0.75$

The two responses in each row of Table 5.7 are the values obtained at both levels of B. Since we decided B and its interactions were trivial, we can ignore which level of B gave which response.

The computation of 95% confidence intervals will be referred to later. We can, however, be highly satisfied by their narrowness as we can also be satisfied by the residuals that clearly do not contain a rogue value. Notice that the residuals are somewhat smaller than the 95% confidence intervals – a fact we will return to later.

We can also use the predicted values to produce a diagrammatic representation of the analysis as shown in Figure 5.1.

In drawing the straight lines we should be aware that we have only two points. There could well be pronounced curvature so we know nothing about possible curvature in the middle. However, the main message is that % defectives can be reduced if we use low current and high concentration and further experimentation may be well worthwhile if we examine combinations of current and concentration around 40 A and 32 g/kg.

5.5   Computation of confidence interval

In order to calculate confidence intervals it is first necessary to obtain a measure of the batch-to-batch variation. This may be referred to as standard deviation, combined SD or residual SD,

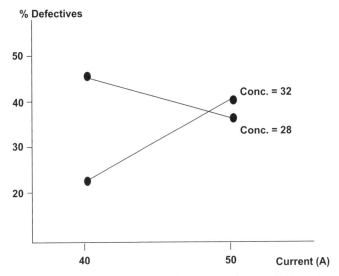

Figure 5.1   Effect of current and concentration on % defectives.

depending upon its method of computation. Whatever its title, its main feature is that it must be a relevant measure of the batch-to-batch SD.

In Chapter 4 we referred to three ways of obtaining an estimate of the batch-to-batch SD:

1. From the experiment itself. We will return to this later in this chapter.

2. Producing replicate batches at one or more set of conditions.

3. By using past production records.

Jones decides to use the third method. A record of 30 batches produced under similar conditions gave a SD of 1.4 based on 29 degrees of freedom.

5.6   95% Confidence interval for an effect

$$\text{Effect} \pm \frac{2ts}{\sqrt{n}}$$

where s is the estimate of SD (1.4 based on 29 degrees of freedom),

$t = 2.04$ (From Table A.2 at a 5% significance level with 29 degrees of freedom)

and n is the number of observations in the experiment ($n = 8$).

The confidence interval is therefore:

$$\text{Effect} \pm \frac{2 \times 2.04 \times 1.4}{\sqrt{8}}$$

i.e.   Effect ± 2.02

5.7 95% Confidence interval for a predicted value

$$\text{Predicted value} \pm ts\sqrt{\frac{q+1}{n}}$$

where q is the number of effects used to compute the predicted value. We used A, C and AC and therefore $q = 3$. The confidence interval is therefore:

$$\text{Predicted value} \pm ts\sqrt{\frac{4}{8}}$$

$$\text{Predicted value} \pm 2.02$$

5.8 Sequencing of the trials

During the discussion of the experimental design we have presented all the data in the order of the standard design matrix. This makes it easy to see the symmetry of the factorial design but it is certainly not the order in which we would carry out the experiment. If there was a drift or a step change in the process during the course of the experiment, and if the batches had been produced in the systematic order of the design matrix, it would be impossible to decide whether effects found were genuine or could have been due to a process change. Experimenters who are unaware of this ambiguity could well draw completely erroneous conclusions.

It is therefore important to sequence the trials carefully so that we minimize the possibility of making wrong inferences from them.

One way of sequencing the trials is to choose the order randomly; sometimes a random order can, by chance, coincide with one of the effects we wish to estimate, so inspection of the random order is still required and if necessary another sequence using random numbers will have to be obtained. There are, however, better ways of sequencing factorial experiments that will be discussed in later chapters.

5.9 Were the aims of the experiment achieved?

Yes.

Jones has established that the current and tin concentration, but not the bath temperature, affect the per cent defectives. He discovered that there was an interaction between the current and the concentration.

He concluded that a considerable decrease in per cent defectives could be achieved with low current (40 A) and high tin concentration (32 g/kg), significantly lower than any of the other conditions tried in the experiment. This should lower the defective rate from 45% to around 22%.

But..

Jones has established the best of the four conditions, but he has not minimized the per cent defectives, for which he would need to undertake further experimentation in the region of the best conditions so far.

Bath temperature was not significant, but only between the two temperatures chosen. A wider range might have led to a significant difference.

We do not know in what order the trials were done. Did he follow the advice in the previous section?

Did he include all the important variables in the process? Were there others that might have had an effect?

Lastly, but most importantly, reducing the defective rate to 22% is good but still not acceptable. There seems to be an endemic problem in the company that they are prepared to obtain good quality by 100% inspection rather than building new processes. A culture change is clearly needed!

5.10   A four-variable experiment

Jones was reasonably satisfied with the outcome of his experiment. A colleague has questioned whether he did in fact include all the important variables and suggests that the dwell-time (the time that the transistors are in the coating bath) may have some effect on the thickness and thus on the defective rate.

Jones therefore decides to extend his original 8 trials by adding dwell-time (D) to the experiment. The first 8 trials used a dwell-time of 8 min, Jones proposes to use a dwell-time of 10 min for the next 8 trials that will use the same combinations of A, B and C.

He will aim to determine the effect of changing dwell-time from 8 min to 10 min on the defective rate, confirm the effects of variables A, B and C over the larger experiment, identify any further interactions that may occur, and determine the conditions that will minimize the per cent defectives.

The full design together with the responses is given in Table 5.8.

Table 5.8   A 2^4 experiment

A	B	C	D	% defectives
−	−	−	−	46.3
+	−	−	−	36.2
−	+	−	−	44.1
+	+	−	−	36.4
−	−	+	−	21.4
+	−	+	−	40.8
−	+	+	−	22.7
+	+	+	−	39.3
−	−	−	+	41.7
+	−	−	+	33.4
−	+	−	+	41.2
+	+	−	+	34.8
−	−	+	+	18.9
+	−	+	+	36.2
−	+	+	+	19.7
+	+	+	+	37.0

From Table 5.8 we can obtain estimates for

4 main effects;
6 two-variable interactions;
4 three-variable interactions;
1 four-variable interaction.

It would be surprising if all the two-variable interactions were significant. It would be even more surprising if the three-variable interactions or the four-variable interactions were significant. In fact some of these interactions may have no physical meaning. Surely we must be able to estimate the residual standard deviation from such a large experiment in which many of the effects are expected to be nonsignificant.

All the estimated effects, calculated using the design matrix method introduced earlier in the chapter, are given in Table 5.9.

Table 5.9 Estimated effects for the 2^4 experiment

Effect	Estimate	Effect	Estimate
A	4.76	D	−3.04
B	0.04	AD	0.21
AB	0.19	BD	0.59
C	−9.76	ABD	0.29
AC	12.89	CD	−0.06
BC	0.31	ACD	−0.56
ABC	−0.89	BCD	−0.14
		ABCD	0.41

5.11 Half-normal plots

One powerful diagnostic technique for factorial experiments is a half-normal plot. In order to carry out a plot we arrange the effects (15 in this experiment) in ascending order of magnitude (ignoring the sign) and display them on a half-normal probability plot. As its name suggests, the half-normal probability plot uses half the scale of the ordinary Normal probability plot starting from the mid-point of the normal distribution. Also, for ease of plotting the probabilities are replaced by the ranks. The horizontal positions of the ranked effects are obtained from Table A.9.

Effects that are comprised only of random variation should approximately be in a straight line going through the origin.

Effects that are significant will not conform to this linearity.

The lowest effect is 0.04 for B, the next lowest in magnitude is *CD* whose effect was −0.06, and so on up to the highest effect, 12.89 for *AC*.

Figure 5.2 shows the half-normal plot for the effects given in Table 5.9.

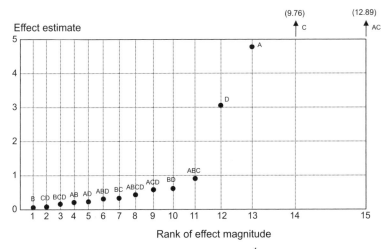

Figure 5.2   Half-normal plot of 2^4 experiment.

The values for the highest effects, C (9.76) and AC (12.89), have been excluded from the plot since their magnitudes would require a considerably extended scale.

In real life, points do not usually fall exactly on a straight line so some judgement is necessary. There can, however, be little doubt that variable D is not due to random error so any effect of greater magnitude – A, C and AC – must also be excluded from estimating the random error. However, a strong case could be made for including the other 11 points and we shall use this point to obtain an estimate of the residual standard deviation. This is achieved using the following formula:

$$\text{RSD} = \sqrt{\frac{n}{4} \frac{\sum (\text{Relevant effect estimates})^2}{\text{Number of effects}}}$$

Note: the formula refers **only** to those effects (the 11 'relevant' ones that were considered to be on the straight line though the origin) included in the computation of the RSD.

In the formula n is the number of trials in the experiment (in this case 16) and '4' is fixed.

$$\text{RSD} = \sqrt{\frac{16}{4} \frac{[(0.04)^2 + (0.19)^2 + \ldots + (0.41)^2]}{11}}$$

$$= \sqrt{4 \times \frac{1.9071}{11}}$$

$$= 0.83 \text{ based on 11 degrees of freedom}$$

The number of degrees of freedom is 11, because we are combining 11 estimates of the variability each having one degree of freedom.

We can now use this figure to obtain confidence intervals.

$$\text{Width of 95\% confidence interval} = \pm\frac{2ts}{\sqrt{n}}$$

(t based on 11 degrees of freedom $= 2.20$)

$$= \pm\frac{2 \times 2.20 \times 0.83}{\sqrt{16}}$$

$$= \pm 0.91$$

Thus, 4 effects are significant:

$$
\begin{array}{rl}
A: & 4.76 \pm 0.91 \\
C: & -9.76 \pm 0.91 \\
AC: & 12.89 \pm 0.91 \\
D: & -3.04 \pm 0.91
\end{array}
$$

The complete analysis matrix for a 2^4 experiment, in standard design order, is shown in Table A.8. This includes the designs for 2^2 and 2^3.

5.12   Were the aims of the experiment achieved?

Yes.

Jones has confirmed the effects of the current and tin concentration and their interaction, and the lack of effect of the bath temperature within the range investigated.

He appears to have discovered an effect of the dwell-time. A longer dwell-time has reduced the per cent defectives.

But..

Jones is slightly disappointed with the outcome of the extended experiment. True, he has found the effect of D (dwell-time) to be significant, but he is fully aware that the sequencing of the trials was not satisfactory as the 8 trials at the higher dwell-time sequenced together, following the first set of 8 trials at the lower dwell time; the effect he has found may be due to a difference in per cent defective between the two groups of trials and he will need further work to check on dwell time.

Increasing the dwell-time to 10 min should reduce the defective rate from 22% to 19%. Although this will be an improvement it is still unsatisfactory. There is also a down side to this change since increasing the dwell-time may result in reduced capacity.

5.13   Problems

1. Dullness has been observed in recent batches of a particular dyestuff and Dr Stayfast has designed a laboratory experiment in order to discover the cause of the dullness. The last 10 batches that have been produced on the plant have brightness values in the range

3 to 7 with an average brightness of 5. It is important that the plant operating conditions be changed as soon as possible to obtain a higher brightness. (Brightness is measured on an instrument that gives a digital display from 0 to 14.)

It is decided that 3 variables will be included in the laboratory experiment, the details of which are tabulated below:

Trial number	Speed of agitation (A)	Reaction temperature (B)	pH (C)	Brightness
1	Fast	80°	6	5
2	Fast	90°	8	7
3	Slow	90°	6	5
4	Fast	80°	8	9
5	Slow	80°	8	3
6	Slow	80°	6	7
7	Fast	90°	6	1
8	Slow	90°	8	3

(a) The trials have been recorded in the order in which they were undertaken. Using the factorial analysis program, enter the brightness values in the appropriate places corresponding to the levels of the variables used, in the order of the standard form of the design matrix.

(b) Obtain estimates for the main effects and interactions.

(c) Using a half-normal plot, decide which effects are significant.

(d) Four trials were previously carried out under the same conditions – medium speed, $85\,°C$, pH of 7 – and gave a standard deviation of 0.80. Use this value as an estimate of the residual standard deviation to obtain a confidence interval for the effect estimates and hence decide which effects are significant. Are your conclusions the same as when using the half-normal plot?

2. Dr Stayfast decides to carry out a larger experiment to confirm his conclusions and assess the effect of a further variable, catalyst concentration (D), on brightness. He also decides to use the mean of visual assessments by a panel of 5 inspectors. The results of his experiment, in the order of the standard design matrix, are:

6.8, 6.8, 7.2, 4.4, 5.0, 9.4, 4.8, 9.0, 6.8, 3.8, 5.4, 2.6, 3.0, 8.2, 2.8, 8.6.

Using the factorial analysis program, enter the data.

(a) Determine the main effects and interactions.

(b) Using the half-normal plot, decide which effects are significant.

(c) Compare the conclusions with those of the experiment in Problem 1.

3. In Problem 2 of Chapter 4 Dr. Scratchplan, a research scientist at a chemical plant, was conducting a series of experiments to determine conditions that would improve the plasticity of a polymer. He was unwilling to explore the complete range of the variables

until he had made a preliminary investigation over a narrower range around the present operating conditions.

He decided to use initial values as follows:

Pressure (P) 1.3 and 1.6 atm
Temperature (T) 271°C and 279°C
Agitation speed (S) 50 and 60 r.p.m.

In his first preliminary investigation he decided to do a 2^2 factorial experiment varying only two of the possible factors, pressure and temperature. To obtain an estimate of the error he decided to duplicate each of the four trials.

His results were as follows:

P	T	Plasticity	Plasticity	
			$\bar{x}$	s
1.3	271	213		
1.3	271	225	219	8.5
1.3	279	257		
1.3	279	245	251	8.5
1.6	271	258		
1.6	271	266	262	5.7
1.6	279	281		
1.6	279	289	285	5.7

The pattern of speeds he used with the combinations of temperature and pressure was:

T	P	
	1.3	1.6
271	50 r.p.m.	60 r.p.m.
279	60 r.p.m.	50 r.p.m.

On examination, Dr Scratchplan realizes the folly of his design. Trials carried out with an agitation speed of 50 also correspond to the high level of the interaction (pressure × temperature) while an agitation speed of 60 corresponds to the low level. Thus the design cannot discriminate between any effects due to agitation speed and those due to the interaction (pressure × temperature).

Because of this confusion he decides to do 4 further trials in duplicate so as to obtain a full factorial design over the 2 experiments.

The additional experimental results turn out to be:

T			P		
	1.3			1.6	
271	236	$\bar{x} = 228$	286		$\bar{x} = 281$
	220	$s = 11.3$	276		$s = 7.1$
	60 r.p.m.			50 r.p.m.	
279	275	$\bar{x} = 268$	286		$\bar{x} = 283$
	261	$s = 9.9$	280		$s = 4.2$
	50 r.p.m.			60 r.p.m.	

(a) Using the eight means as your response, enter them into the program and determine the main effects and interactions.
(b) The residual SD, using the standard deviations of each pair of replicates, is 7.93 with 8 degrees of freedom. Calculate the width of confidence interval for any effect.
(c) Interpret any significant effects found.

6

More for even less: using a fraction of a full design

6.1 Introduction

In Chapters 4 and 5 we have seen the benefits of factorial experiments. They allow us to evaluate simultaneously the effects of several variables and all their interactions. The design is so chosen to be balanced in each variable and in every combination of variables, and has the same efficiency for each comparison as a similarly sized experiment to investigate only one variable.

With many variables, however, the experiment becomes unwieldy and yields far more information than is desired. For example, a two-level factorial experiment with six variables would require 64 trials and would lead to the estimation of no fewer than 63 effects:

6 main effects	(A, B, C, D, E, F)
15 two-variable interactions	$(AB, AC$, etc.)
20 three-variable interactions	$(ABC, ABD$, etc.)
15 four-variable interactions	$(ABCD, ABCE$, etc.)
6 five-variable interactions	$(ABCDE, ABCDF$, etc.)
1 six-variable interaction	$(ABCDEF)$

It is very doubtful if **all** of these estimates would be required. In practice, interactions involving more than two variables are unlikely to be meaningful.

Since the 2^6 experiment is, on the one hand, too large and, on the other hand, too productive of effect estimates it is tempting to ask 'Can I carry out **part** of the 2^6 factorial experiment in order to get the estimates I require while losing those that I neither need nor understand?' The answer is 'Yes,' and we can use the methods developed in this chapter to help the researcher to select **which** part of the 2^6 factorial experiment he will carry out.

Effective Experimentation: For Scientists and Technologists   Richard Boddy and Gordon Smith
© 2010 John Wiley & Sons, Ltd

6.2   Obtaining half-fractional designs

To illustrate the method of obtaining a fractional design we shall look at a $\frac{1}{2}(2^3)$. Nobody suggests that a $\frac{1}{2}(2^3)$ is a worthwhile design, but it is presented only for illustration purposes since it has the advantage of being easy to follow.

Let us look at an example from the world of sport rather than industry. Wheelwright, a professional cyclist, is to race on a one-mile lap around the town circuit. He believes that the race will be decided by a last lap sprint and wishes to select the best combinations of settings for the fastest speed. Initially he selects three variables:

Seat height	A
Handlebar height	B
Tyre pressure	C

The response is the time in seconds for a single lap.

He decides to carry out only four trials of the eight in a full factorial design. He selects the four using random numbers and obtains the responses shown in Table 6.1.

Table 6.1   Three-way table for a half-fraction of the 2^3 experiment

A	Low B		High B	
	Low C	High C	Low C	High C
Low		123.2	123.6	
High		127.2		128.4

Clearly there is a lack of balance in Table 6.1. This lack of balance is illustrated by computing an analysis matrix as given in Table 6.2.

Table 6.2   Analysis matrix for a half-fraction of the 2^3 experiment

A	B	AB	C	AC	BC	ABC	y
−	+	−	−	+	−	+	123.6
+	+	+	+	+	+	+	128.4
+	−	−	+	+	−	−	127.2
−	−	+	+	−	−	+	123.2

When we come to estimate the main effect C, we notice that the C column contains 3 pluses and only 1 minus. Clearly this particular half fraction **is not balanced** unlike the full 2^3 factorial experiment from which it was taken. Whereas each column in the standard design matrix for a 2^3 experiment contains 4 pluses and 4 minuses, there is no such balance of plus and minus signs in Table 6.2. This does not prevent us from calculating effect estimates, but it does mean that we cannot use simplified formulae similar to those that are applicable to the full factorial experiment. With an unbalanced design we would have to work from first principles in calculating effect estimates and sums of squares. Furthermore, we would get less precise estimates than could be obtained from a balanced design of the same size and these estimates would not be independent because of intercorrelations between the columns in the design matrix.

If this particular half-fraction is lacking in the fine qualities that we found in the full factorial experiment, can we find a different half-fraction that is more desirable? Thinking in terms of a balanced design, can we find four rows in the standard analysis matrix for a 2^3 such that each of the seven columns will contain 2 plus and 2 minus signs? Unfortunately, this is not possible. Indeed, it would be rather optimistic to expect **all** the benefits of a 2^3 design from only four trials. Perhaps we should concentrate on getting 'balanced' estimates of the 3 main effects. We can see in Table 6.2 the nature of the imbalance of this particular half-fraction. While the four response values are equally shared between the two rows they are not equally shared among the four columns. If we changed **one** of the four trials a more balanced design would result.

The half-replicate in Tables 6.3 and 6.4 below is certainly better balanced than the one we have just considered. For each of the variables there are two trials at the low level and two at the high level.

Table 6.3 Analysis matrix for a second half-fraction

A	B	AB	C	AC	BC	ABC	y
$-$	$+$	$-$	$-$	$+$	$-$	$+$	123.6
$+$	$+$	$+$	$+$	$+$	$+$	$+$	128.4
$+$	$-$	$-$	$+$	$+$	$-$	$-$	127.2
$-$	$-$	$+$	$-$	$+$	$+$	$-$	124.8

Table 6.4 Three-way table for a second half-fraction

A	Low B		High B	
	Low C	High C	Low C	High C
Low	124.8		123.6	
High		127.2		128.4

From this half-fraction we can calculate the following estimates, using either the design matrix or the three-way table:

$$\text{Main effect } A = 3.6$$
$$\text{Main effect } B = 0$$
$$\text{Main effect } C = 3.6$$
$$\text{Interaction } AB = 1.2$$
$$\text{Interaction } AC = ??$$
$$\text{Interaction } BC = 1.2$$
$$\text{Interaction } ABC = 0$$

You will notice in the above list that there is no estimate for the interaction AC. It is not possible to calculate an estimate since the AC column of the analysis matrix contains four plus signs. You will also notice that the six estimates we **have** calculated fall into three pairs, i.e. we have the same value for main effect A and for main effect C, etc. This is not just a coincidence as Table 6.4 shows. In the design matrix the A column and the C column are **identical**. Similarly, the B column and the ABC column are identical.

Since we have chosen to use only 4 trials of the 8 in a full design, there will always be one column whose signs are all the same and therefore an effect we cannot estimate. Also, 4 trials allow us to estimate only 3 independent effects.

We summarize this situation by saying that for this experimental design:

> AC is the **defining contrast**
> (A and C)
> (B and ABC) are **alias pairs**
> (AB and BC)

The two effects that constitute any alias pair are inseparable from each other. The calculated estimate of 3.6 s for the (A, C) alias pair tells us that a change in time of 3.6 s can be attributed to the change made to seat height **or** to the change made to tyre pressure **or** to both. As you can see in Table 6.3, the A column and the C column are identical. An experiment in which a change in one variable is **always** accompanied by a change in a second variable will never allow us to separate the effects of the two variables.

Since the least important of all the seven estimates is probably the interaction ABC, we will choose a half fraction that has the interaction ABC as the defining contrast. To do this we refer to the standard analysis matrix and select the four trials that have a plus sign in the ABC column. This third example of a half-fraction is set out in Tables 6.5 and 6.6.

Table 6.5 Analysis matrix for a third half-fraction

A	B	AB	C	AC	BC	ABC	y
$-$	$+$	$-$	$-$	$+$	$-$	$+$	123.6
$+$	$+$	$+$	$+$	$+$	$+$	$+$	128.4
$-$	$-$	$+$	$+$	$-$	$-$	$+$	123.2
$+$	$-$	$-$	$-$	$-$	$+$	$+$	128.0

Table 6.6 Three-way table for a third half-fraction

A	Low B		High B	
	Low C	High C	Low C	High C
Low		123.2	123.6	
High	128.0			128.4

Without calculating effect estimates we can see which effects will be aliased with each other by spotting pairs of identical columns in the design matrix. In Table 6.5 we see that:

Interaction ABC is the defining contrast,

Main effect A is aliased with interaction BC,

Main effect B is aliased with interaction AC,

Main effect C is aliased with interaction AB.

With this half-fraction then, we could estimate the main effect A **if** we could assume that interaction BC did not exist, we could estimate the main effect B if we could assume that interaction AC did not exist and similarly for C assuming no AB interaction. These assumptions would not be made on statistical grounds, of course. One wonders if there are **any** situations in which such assumptions could reasonably be made and therefore the half-fraction of the 2^3 experiment is not a widely used design. Though we have used a 2^3 factorial experiment to illustrate the ideas underlying fractional factorials, the practical benefits of the technique will be enjoyed by the researcher who wishes to investigate the effect of **4 or more variables**.

Perhaps we should summarize our procedure to date:

(a) To obtain a half-fraction you need to select a defining contrast.

(b) We then select from the full factorial analysis matrix all the trials with a '+' in the column for the defining contrast.

(c) We then compare the other columns, in the half-fraction, and any columns that are the same are classed as alias pairs.
Note that if the defining contrast is one with all negative signs, every sign will be reversed in the columns corresponding to an alias pair.

(d) An alias pair must only contain, at the most, **one** effect of interest.

If we have two effects that are of interest in the same alias pair we will not be able to separate them; we should repeat the procedure with another defining contrast to find a suitable one. The $\frac{1}{2}(2^3)$ was clearly unsatisfactory with respect to separating the interactions and main effects. We shall go on to design a $\frac{1}{2}(2^4)$ but, before we do so, let us look more closely at the defining contrasts for the previous designs.

6.2.1   With defining contrast ABC

Alias pairs (A, BC) (B, AC) (C, AB). Multiplying the effects in the alias pairs gives (ABC) (ABC) (ABC), in other words the defining contrast.

6.2.2   With defining contrast AC

Alias pairs (A, C) (B, ABC) (AB, BC). Multiplying the effects in the alias pairs gives (AC), (AB^2C), (AB^2C). If we define a squared term as equal to '1' we again obtain the defining contrast in every instance.

We shall now use, to our advantage, the relationship between alias pairs and the defining contrast.

6.3 Design of $^1/_2(2^4)$ experiment

Wheelwright decides to introduce a fourth variable – type of tyre (D) – into his experiment. There are two types he wishes to investigate: X147 and X591. We see that this is a qualitative variable rather than the measured variables we have previously used. However, a qualitative variable with **only two** types can easily be incorporated into a factorial design. We code X147 as the low level and X591 as the high level.

Wheelwright also decides that he is mainly interested in the following (stipulated) effects:

$$\begin{array}{ll} \text{Main effects} & A \ B \ C \ D \\ \text{Interactions} & AB \ BC \end{array}$$

We shall not look behind Wheelwright's reasons for his lack of interest in the other interactions. Perhaps he has knowledge from previous experience.

Notice that he could have estimated up to seven effects from eight trials, thereby using all the degrees of freedom. This is not good. Some must be left to estimate the residual.

We shall now follow a procedure that will, in most situations, enable us to obtain a half-fraction.

(a) Write down the effects for a full experiment with 8 trials – that is a 2^3:

$$A \quad B \quad AB \quad C \quad AC \quad BC \quad ABC$$

Indicate with asterisks those effects that are stipulated:

$$A* \quad B* \quad AB* \quad C* \quad AC \quad BC* \quad ABC$$

(b) Set D as an alias of one of the nonstipulated effects, for example set D as an alias of ABC. This gives $ABCD$ as defining contrast.

(c) Multiply each effect in (a) by $ABCD$, replacing any squared letter by 1, e.g. $A \times ABCD = A^2BCD = BCD$.

The resulting alias pairs are as shown in Table 6.7:

Table 6.7 Alias pairs for a $^1/_2(2^4)$ design with $ABCD$ as defining contrast

Alias pairs
$A*$, BCD
$B*$, ACD
$AB*$, CD
$C*$, ABD
AC, BD
$BC*$, AD
ABC, $D*$

Since all of the stipulated effects are unconfounded with each other (that is, no two stipulated effects are in the same alias pair), we have succeeded in identifying a suitable design.

If two stipulated effects were in the same pair (that is, they were confounded) it would not be possible to estimate them separately and a different design would have to be obtained.

We should note that $D = AC$ also forms a suitable design. Which of the two designs is more suitable will depend upon other considerations, but if these are nonexistent it is usual to take the one with the highest number of variables in the defining contrast, i.e. $D = ABC$.

We can easily obtain the design matrix for the $\frac{1}{2}(2^4)$ experiment by first writing down the design matrix for a 2^3 experiment with factors A, B and C and then generating the column for D. Having decided to use $ABCD$ as the defining contrast, we know that the D column will be identical to the ABC column. The design matrix is shown in Table 6.8 along with the response (the time for each trial).

Table 6.8 Design matrix and responses for $\frac{1}{2}(2^4)$ experiment

A	B	C	$D = ABC$	Time
$-$	$-$	$-$	$-$	122.3
$+$	$-$	$-$	$+$	129.4
$-$	$+$	$-$	$+$	126.1
$+$	$+$	$-$	$-$	126.0
$-$	$-$	$+$	$+$	122.1
$+$	$-$	$+$	$-$	132.0
$-$	$+$	$+$	$-$	129.5
$+$	$+$	$+$	$+$	125.9

6.4   Analysing a fractional experiment

We can obtain the effects using the design matrix for a full 2^3 design, with the alias written down with each effect.

The full list of effects is given in Table 6.9.

Table 6.9 Effect estimates for a $\frac{1}{2}(2^4)$ experiment

Effect (alias pair)	Effect estimate
A, BCD	3.32
B, ACD	0.43
AB, CD	-5.17
C, ABD	1.43
AC, BD	-0.17
BC, AD	0.23
ABC, D	-1.58

The largest effects are A (or BCD) and AB (or CD). To determine whether or not they are significant we need to obtain confidence intervals based on an external estimate of the residual standard deviation or on an estimate derived from a half-normal plot. We shall use a half-normal plot shown in Figure 6.1.

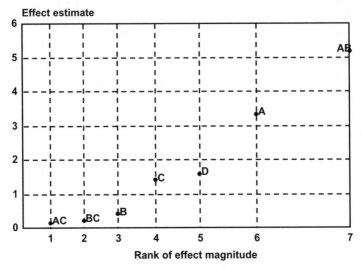

Figure 6.1   Half-normal plot of $\frac{1}{2}(2^4)$ experiment.

From an inspection of the half-normal plot, the five lowest effects should be considered as residual error. From these we obtain an estimate of the residual standard deviation:

$$\text{RSD} = 1.38 \text{ with 5 degrees of freedom}$$

95% confidence interval for each effect

$$= \pm \frac{2ts}{\sqrt{n}}$$

t (from Table A.2 at the 95% confidence level with 5 degrees of freedom) $= 2.57$
n, the number of trials in the experiment $= 8$

The confidence interval is therefore

$$\pm \frac{2 \times 2.57 \times 1.38}{\sqrt{8}}$$

i.e. ± 2.5

We can therefore decide that only the main effect A and the interaction AB are significant. The average times at each combination of levels of A and B are given in Table 6.10.

The recommended conditions would appear to be low seat height and low handlebar height, with conditions of tyre pressure and tyre type immaterial.

It may not be as simple as this. AB, the interaction between seat height and handlebar height, was a stipulated effect; Wheelwright also stipulated BC but was not interested in any other interactions. It is possible that he may have been wrong in his assumptions and CD (the alias partner of AB, representing the interaction between tyre pressure and tyre type) might exist.

Table 6.10 Mean times and levels of A and B

		Seat height (A)	
		Low	High
Handlebar height (B)	Low	122.2	130.8
	High	127.8	126.0

The significant interaction could be AB, or it might be CD, or a combination of both. When we use a fractional factorial experiment, we must remember that we reduce the number of effects we can estimate and our conclusions are crucially dependent upon the assumptions that we use in selecting our stipulated effects.

6.5 Summary

We now see how to obtain the designs for fractional factorial experiments. The approach can be extended to 16 or more trials in an experiment. We will not pretend this is easy and it is far more likely that you will select a design from a compendium like the section at the end of this chapter or from a computer program. However, this is dangerous unless we take forward the knowledge gained in this chapter. The most important feature of any factorial experiment is:

Which are the alias pairs?

With larger fractions the aliases will be in groups rather than pairs. For example with a $\frac{1}{4}$ fraction there will be four aliases in a group.

The second question is:

How do we find alias groups?

(i) Using the defining contrast(s) as shown in this chapter. With higher fractions there is more than one defining contrast. For example with a $\frac{1}{4}$ fraction there are 3 defining contrasts.

(ii) Using a computer program to obtain correlations among the independent variables and their interactions (see Chapter 9) including those of higher order.

One useful point to note is generally to limit the aliases to two-variable interactions only since it is unlikely that higher-order interactions are of interest.

Having obtained the aliases it is important to recognize them when analysing the data. Good practice is to write in all aliases alongside the magnitude of each effect but underline the stipulated main effect or interaction.

6.6 Did Wheelwright achieve the aims of his experiment?

Yes.

He was able to determine a setting that would give him a time round the circuit that was faster than other settings, and decide whether it was significantly better.

He was able to do this with a small number of circuits, using 8 of the maximum 16 settings. He had a measure of run-to-run variability.

But...

The conclusion had some doubt, as he had made assumptions about that interactions would occur. He found that the seat height × handlebar height interaction was significant, but because of the fractional design it could equally have been tyre pressure and tyre type. We do not know the strength of his evidence for rejecting many interactions when he designed the experiment. If both these interactions do exist, he needs to design further trials to sort out this ambiguity.

His measure of run-to-run variability was estimated from the factors which appeared to represent only random variation. If he required a good measure he would need to replicate two or three conditions or use past data.

6.7   When and where to choose a fractional design

We have seen that the fractional factorial experiment allows the information of interest (main effects and some two-factor interactions) to be obtained with a fraction of the effort required for a full factorial experiment, while not being burdened by estimates of higher-order interactions.

As information is sacrificed, there is a danger that ambiguous conclusions may be obtained. What are the situations for which fractional factorial experiments are or are not recommended?

6.7.1   Three variables

A half fraction, $\frac{1}{2}(2^3)$ experiment, is conducted in 4 trials, so no more than 3 effects can be estimated.

Only the 3 main effects A, B and C can be obtained. The defining contrast is ABC. Each main effect is aliased with one of the interactions. It must be assumed therefore that interactions do not exist.

The design is shown in Table 6.11.

Table 6.11 A $\frac{1}{2}(2^3)$ design

A	B	C
−	−	−
+	−	+
−	+	+
+	+	−

A fractional factorial experiment in three variables is not recommended except as part of a greater design.

6.7.2   Four variables

A half fraction, $\frac{1}{2}(2^4)$ experiment, is conducted in 8 trials, so no more than 7 effects can be estimated.

These are usually the 4 main effects and up to 3 two-factor interactions. However, it does not work for just any set of 3 interactions.

If the three stipulated interactions have one factor in common (assuming $ABCD$ is the defining contrast), e.g. they are AB, AC and AD, they can be kept in separate alias pairs from each other and from the main effects.

A $\frac{1}{2}(2^4)$ experiment with defining contrast $ABCD$ is shown in Table 6.12.

Table 6.12   A $\frac{1}{2}(2^4)$ design

A	B	C	D
−	−	−	−
+	−	−	+
−	+	−	+
+	+	−	−
−	−	+	+
+	−	+	−
−	+	+	−
+	+	+	+

If BC, BD and CD are negligible we can obtain unbiased estimates of all main effects and interactions AB, AC and AD.

If two stipulated interactions cover all four letters (e.g. AB and CD) then they will be in the same alias group and cannot be separately estimated. A different defining contrast (e.g. ABC) will not help, as it will lead to one of these interactions being confounded with a main effect.

If such interactions are likely to be of interest, it would be better to use a full factorial.

6.7.3   Five variables or more

With five variables or more, a half-fraction permits all main effects and two-factor interactions to be separately estimated.

With five variables and defining contrast $ABCDE$ the 5 main effects and 10 two-factor interactions can be kept in different alias groups.

This design is shown in Table 6.13.
This design enables all main effects and all 2-factor interactions to be estimated.

A quarter-fraction in five variables, $\frac{1}{4}(2^5)$ in 8 trials, will yield estimates of the 5 main effects and only 2 two-factor interactions.

The design is shown in Table 6.14.

Providing all interactions except AB and AD are negligible we can obtain unbiased estimates of all main effects and interactions AB and AD. The presence of any other interactions will cause confusion. Note also that quarter-fractions result in alias groups of four, which may include one main effect and 3 interactions of which more than one may be two-factor interactions. It would be better to go up to a half-fraction as shown in Table 6.13.

Table 6.13 A $\frac{1}{2}(2^5)$ design

A	B	C	D	E
−	−	−	−	+
+	−	−	−	−
−	+	−	−	−
+	+	−	−	+
−	−	+	−	−
+	−	+	−	+
−	+	+	−	+
+	+	+	−	−
−	−	−	+	−
+	−	−	+	+
−	+	−	+	+
+	+	−	+	−
−	−	+	+	+
+	−	+	+	−
−	+	+	+	−
+	+	+	+	+

Table 6.14 A $\frac{1}{4}(2^5)$ design

Defining contrasts: *ABCD*, *ACE* and *BDE*				
A	B	C	D	E
−	−	−	−	+
+	−	−	+	−
−	+	−	+	+
+	+	−	−	−
−	−	+	+	−
+	−	+	+	+
−	+	+	+	−
+	+	+	+	+

A quarter-fraction in six variables permits 6 main effects and 9 interactions to be estimated. Depending on the choice of defining contrast, only 7 of the interactions will be two-factor ones. It is unlikely that as many interactions may be of interest, so this should be a suitable design. The design is shown in Table 6.15.

Providing interactions *BD*, *CD*, *CE*, *CF*, *DE*, *DF* and *EF* are negligible we can obtain unbiased estimates of all main effects and interactions *AB*, *AC*, *AD*, *AE*, *AF*, *BC* and *BE*.

Even more economical designs may be used with higher numbers of variables. In Chapter 7 we shall meet a design that can estimate up to 7 main effects with only 8 trials.

Table 6.15 A $\frac{1}{4}(2^6)$ design

Defining contrasts: *ABDF*, *ACDE* and *BCEF*

A	B	C	D	E	F
−	−	−	−	−	−
+	−	−	−	+	+
−	+	−	−	−	+
+	+	−	−	+	−
−	−	+	−	+	−
+	−	+	−	−	+
−	+	+	−	+	+
+	+	+	−	−	−
−	−	−	+	+	+
+	−	−	+	−	−
−	+	−	+	+	−
+	+	−	+	−	+
−	−	+	+	−	+
+	−	+	+	+	−
−	+	+	+	−	−
+	+	+	+	+	+

6.8 Problems

1. A researcher is investigating the effect of production methods on the refractive index of glass. He has identified 4 variables that he would like to investigate, namely:

   Recycle feed rate (*A*):   low or high
   Pressure (*B*):   low or high
   Evaporation level (*C*):   low or high
   Steam rate (*D*):   low or high

   He wishes to evaluate the four main effects and the following interactions:

   Recycle feed rate   × Pressure
   Pressure   × Evaporation level
   Pressure   × Steam rate

   He is prepared to assume that all other interactions are negligible. Design an experiment with 8 trials to investigate the stipulated effects.

2. In Problem 1 the researcher wished to design a $\frac{1}{2}(2^4)$ experiment to investigate four variables. He eventually decided to use *ABCD* as the defining contrast and obtained the responses that are listed below together with the design matrix:

Recycle feed rate A	Pressure B	Evaporation level C	Steam rate D	Refractive index
−	−	−	−	83
+	−	−	+	111
−	+	−	+	141
+	+	−	−	85
−	−	+	+	107
+	−	+	−	91
−	+	+	−	82
+	+	+	+	138

The value of the residual standard deviation in a previous experiment was 4.1 based on 7 degrees of freedom. Evaluate the significance of the 4 main effects and the three stipulated interactions and thus decide the conditions that should be used to maximize the refractive index.

7

Saturated designs

7.1   Introduction

We have previously examined two-level factorial designs. We observed their great efficiency in determining main effects and interactions. However, when investigating many variables, a full factorial design requires too many trials and provides an excess of information so we need a suitable fraction of a full design.

The most economical of fractional designs is a saturated design in which many variables are investigated in the minimal number of trials. For example, a saturated design can investigate 11 variables in 12 trials (the maximum number of variables is one less than the number of trials). However, it is not always wise to use a fully saturated design in case there are operational problems with the experiment that lead to the failure to complete one (or more) trials. Such a failure could result in the failure of the whole experiment. In the case we shall look at in this chapter seven variables are investigated in 12 trials, a far more realistic number.

The saturated designs are highly efficient at investigating many variables but this efficiency is obtained at a great cost – the inability to investigate interactions. Thus, if interactions are suspected to be dominant, these designs are not advisable.

Saturated designs come under many names but they all give the same designs. In this chapter we shall look at a Plackett–Burman design, but others include Taguchi designs and, more generally, fractional factorials.

7.2   Towards a better oil?

Doug Hoyle of Hoyle Ltd is developing a new oil to give lower fuel consumption (as measured by litres per 100 km) than conventional oil. From the experience of the research and development department of Hoyle Ltd there are believed to be seven major components in the formulations. These are:

Additive package (A_1 or A_2): A_2 is a new package developed by Parisbus Ltd and is claimed to lower fuel consumption.

Effective Experimentation: For Scientists and Technologists   Richard Boddy and Gordon Smith
© 2010 John Wiley & Sons, Ltd

Base oil (B_1 or B_2): Two types of base oils are available.

Anticorrosive additive (C_1 or C_2): It is believed that an anticorrosive additive should not alter fuel consumption but C_2 is far cheaper than C_1 and Hoyle would wish to include it in order to offset other costs.

Detergent (D_1 or D_2): Detergents are added to clean the cylinders and it is likely that an efficient detergent would decrease fuel consumption.

Ester/Polymer (E_1 or E_2): The oil is partially synthetic and a polymer is included to reduce the effect of temperature on viscosity.

Friction modifier (F_1 or F_2): This is supposed to alter the friction and therefore lower fuel consumption.

Gunk (G_1 or G_2): A special ingredient only known to Doug Hoyle.

The present oil uses $A_1B_1C_1D_1E_1F_1G_1$.

Doug proposes to carry out an experiment to investigate the effects of these components on the fuel economy. Whilst he would hope that each and every component would give a significant decrease in fuel consumption (apart from the new anticorrosive additive that will hopefully not be detrimental), he realizes that it is unrealistic to suppose that this would be the case. The object of running an experiment is to find out which ones should be incorporated.

7.3   The experiment

Doug Hoyle decides to formulate 12 oils and carry out an engine trial on each one. The Plackett–Burman design that he uses, together with the results, is given in Table 7.1. We shall consider how the design was derived later. Let us first look at the principle of the design and then the analysis.

Table 7.1   Oil formulations and results

Formulation	A	B	C	D	E	F	G	Fuel consumption (litres per 100 km)
1	A_2	B_1	C_2	D_1	E_1	F_1	G_2	11.76
2	A_2	B_2	C_1	D_2	E_1	F_1	G_1	11.37
3	A_1	B_2	C_2	D_1	E_2	F_1	G_1	11.91
4	A_2	B_1	C_2	D_2	E_1	F_2	G_1	12.11
5	A_2	B_2	C_1	D_2	E_2	F_1	G_2	11.32
6	A_2	B_2	C_2	D_1	E_2	F_2	G_1	12.02
7	A_1	B_2	C_2	D_2	E_1	F_2	G_2	12.62
8	A_1	B_1	C_2	D_2	E_2	F_1	G_2	12.11
9	A_1	B_1	C_1	D_2	E_2	F_2	G_1	12.40
10	A_2	B_1	C_1	D_1	E_2	F_2	G_2	11.98
11	A_1	B_2	C_1	D_1	E_1	F_2	G_2	12.24
12	A_1	B_1	C_1	D_1	E_1	F_1	G_1	11.74

Before we carry out the analysis it is worthwhile looking at the general design matrix for this experimental design. The design matrix for a saturated design with 12 trials has 11 columns, of which the first ones are used for the variables of interest and the remainder represent residual variation. (It is a special case of a fraction of a 2^{11} design that would require 2048 trials.) The design, with '−' for the low level of a variable and '+' for the high level, is given in Table 7.2. (In the experiment, the current level of each component, with the subscript '1', is considered as the low level.)

Table 7.2 Design matrix for 12 trials

Formulation	Variables							Residual columns				Fuel consumption
	A	B	C	D	E	F	G	H	I	J	K	
1	+	−	+	−	−	−	+	+	+	−	+	11.76
2	+	+	−	+	−	−	−	+	+	+	−	11.37
3	−	+	+	−	+	−	−	−	+	+	+	11.91
4	+	−	+	+	−	+	−	−	−	+	+	12.11
5	+	+	−	+	+	−	+	−	−	−	+	11.32
6	+	+	+	−	+	+	−	+	−	−	−	12.02
7	−	+	+	+	−	+	+	−	+	−	−	12.62
8	−	−	+	+	+	−	+	+	−	+	−	12.11
9	−	−	−	+	+	+	−	+	+	−	+	12.40
10	+	−	−	−	+	+	+	−	+	+	−	11.98
11	−	+	−	−	−	+	+	+	−	+	+	12.24
12	−	−	−	−	−	−	−	−	−	−	−	11.74

We notice the balance of the design. Taking any column we will see there are 6 minuses and 6 pluses. Furthermore, taking any two columns, for example A and B, the six minuses for A are split into 3 minuses and 3 pluses for B. The same is true for the six pluses for A. Thus, the design is perfectly balanced, meaning that every effect can be estimated without any bias from any other effect.

Let us initially concentrate on estimating the effect of each component, using the definition that we first met in Chapter 4.

> *The effect of a component is the expected increase in response as the component moves from its low level to its high level.*

The effect can be estimated from the data. For example, the effect estimate for A is obtained by first averaging the '+' and '−' responses.

The six '+' responses were:

|11.76|11.37|12.11|11.32|12.02|11.98|Mean = 11.76|

The six '−' responses were:

|11.91|12.62|12.11|12.40|12.24|11.74|Mean = 12.17|

The 'effect' of A is the average change from '−' to '+', $11.76 − 12.17 = −0.41$. Thus, changing from additive package A_1 to package A_2 will decrease consumption by 0.41 l/100 km.

We can carry out this procedure for all seven components to give the estimates in Table 7.3.

We can clearly see both the sign and magnitude of each effect estimate. We can also observe that the effects of additive package, anticorrosion component and friction modifier are

Table 7.3 Effect estimates

Component	Code letter	Effect estimate
Additive package	A	−0.41
Base oil	B	−0.10
Anticorrosive additive	C	0.25
Detergent	D	0.05
Ester/polymer	E	−0.02
Friction modifier	F	0.53
Gunk	G	0.08

far higher than the other components. However, to make a more conclusive judgement we need to assess the variability and then decide whether the estimates are due to chance or due to a real effect. To achieve this we use the residual columns, using the fact that they contain a different set of pluses and minuses to divide the data into two halves. The 'effect' of each of these columns, obtained by the difference between the mean of the ' + ' data and the mean of the '−' data, is a random effect due only to the residual variation. We can use their magnitudes to estimate the residual standard deviation, as follows:-

Residual column	Difference between '−' and ' + '	Squared difference
H	0.037	0.001369
I	0.083	0.006889
J	−0.023	0.000529
K	−0.017	0.000289
Sum of squared differences		0.009076

$$\text{Residual SD} = \sqrt{\frac{n}{4}\frac{\sum \text{squared differences}}{No.\text{ of residual columns}}}$$

where n is no. of trials

$$= \sqrt{\frac{12}{4} \times \frac{0.009076}{4}}$$

$$= 0.083 \text{ based on 4 degrees of freedom}$$

This is equivalent to a standard deviation from 5 replicates and we can now use it to assess the effect of each parameter. The 95% confidence interval is given by:

$$\text{effect} \pm \frac{2ts}{\sqrt{n}}$$

From Table A.2 with 4 degrees of freedom at a 95% confidence level $t = 2.78$ giving:

$$= \text{effect} \pm \frac{2 \times 2.78 \times 0.083}{\sqrt{12}}$$

$$= \text{effect} \pm 0.13$$

Thus, three of the parameters have confidence intervals that do not include zero and are therefore significant. They are:-

Additive package (A)	-0.41 ± 0.13
Anticorrosive additive (C)	0.25 ± 0.13
Friction modifier (F)	0.53 ± 0.13

However, only one of these, additive package A_2, decreases fuel consumption from the standard formulation – a considerable disappointment to Doug Hoyle.

7.4   An alternative procedure for estimating the residual SD

The procedure above relies upon several residual columns to estimate the standard deviation. A more practical approach is to use the half-normal plot procedure we met in Chapters 5 and 6. The plot of the magnitudes of the effects, including those of the residual columns, is shown in Figure 7.1.

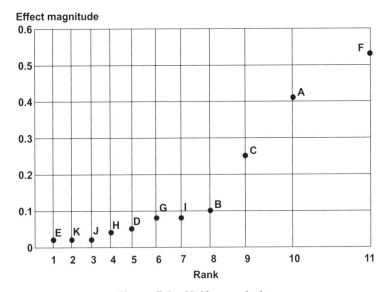

Figure 7.1   Half-normal plot.

First, the plot indicates that there are no outliers, since the smaller effects form a straight line going through the origin. (If there had been an outlier, there would have been a straight line but it would have been offset.) Secondly, it indicates that eight of the columns should be used to calculate the residual. This gives -

Residual SD $= 0.104$ based on 8 degrees of freedom

95% confidence interval for an effect is ± 0.14. This figure is close to the previous estimate and gives support to the conclusions Doug Hoyle drew above.

7.5   Did Doug achieve the aims of his experiment?

Yes.

He explored the effect of seven components of his formulations in an economical experiment with only 12 trials.

He established that three of the seven components could be altered to change the fuel consumption significantly. The new additive package would significantly decrease the fuel consumption, but he was disappointed that by changing two other components, the anticorrosion additive and the friction modifier, made the consumption worse. Perhaps that was good news; apart from the additive A2, the present formulation could not be improved!

He could believe his conclusions because no outliers had distorted the analysis.

But ...

He did not consider the order in which the trials should be undertaken. We shall look at this later.

7.6   How rugged is my method?

Serena Kalm, an analyst at the Steady State Laboratories, is developing an HPLC method and has already carried out a series of experiments to improve the resolution. She now wishes to determine whether the method is robust within the prescribed specification limits for nine parameters. The nine parameters together with the specification limits and experimental limits are given below.

Mobile phase composition (A)	10% ± 1%
pH (B)	6 ± 1
Injection volume (C)	20 ± 5μl
Column temperature (D)	25 ± 1 °C
Flow rate (E)	2.0 ± 0.2 ml/min
Flow temperature (F)	30 °C and 40 °C
Methanol concentration in sample preparation (G)	10 ± 2%
Sonication time (H)	10 ± 2 min
Median particle size (I)	10 μm and 15 μm

To investigate nine parameters, each at two levels, will require a minimum of twelve determinations but it is advisable to increase this to 16 in order to estimate the residual SD and hence the significance of the effects. Serena decides to use 16 trials and uses a standard Plackett–Burman design, shown in Table 7.4.

There are six residual columns shown in Table 7.5:

7.7   Analysis of the design

The design matrix, with conditions represented by a '−' or '+' is shown in Table 7.4, together with the determinations made on the same homogeneous sample.

The effect estimates are shown in Table 7.6.

The residual standard deviation is 0.081 based on 6 degrees of freedom.

The residual standard deviation can be calculated from the residual columns.

Table 7.4 Plackett–Burman design for 16 trials

Trial	Experimental parameters									Resolution
	A	B	C	D	E	F	G	H	I	
1	+	−	−	−	+	−	−	+	+	11.45
2	+	+	−	−	−	+	−	−	+	11.36
3	+	+	+	−	−	−	+	−	−	12.10
4	+	+	+	+	−	−	−	+	−	11.52
5	−	+	+	+	+	−	−	−	+	12.25
6	+	−	+	+	+	+	−	−	−	11.66
7	−	+	−	+	+	+	+	−	−	12.16
8	+	−	+	−	+	+	+	+	−	12.24
9	+	+	−	+	−	+	+	+	+	11.84
10	−	+	+	−	+	−	+	+	+	12.63
11	−	−	+	+	−	+	−	+	+	12.25
12	+	−	−	+	+	−	+	−	+	11.82
13	−	+	−	−	+	+	−	+	−	11.81
14	−	−	+	−	−	+	+	−	+	12.58
15	−	−	−	+	−	−	+	+	−	12.30
16	−	−	−	−	−	−	−	−	−	11.73

Table 7.5 Residual columns

Trial	Residual columns					
	J	K	L	M	N	O
1	−	+	−	+	+	+
2	+	−	+	−	+	+
3	+	+	−	+	−	+
4	−	+	+	−	+	−
5	−	−	+	+	−	+
6	+	−	−	+	+	−
7	−	+	−	−	+	+
8	−	−	+	−	−	+
9	−	−	−	+	−	−
10	+	−	−	−	+	−
11	+	+	−	−	−	+
12	+	+	+	−	−	−
13	+	+	+	+	−	−
14	−	+	+	+	+	−
15	+	−	+	+	+	+
16	−	−	−	−	−	−

Table 7.6 Effect estimates

Parameter	Code Letter	Effect Estimate
Mobile phase composition	A	−0.47
pH	B	−0.04
Injection volume	C	0.34
Column temperature	D	−0.01
Flow rate	E	0.04
Flow temperature	F	0.01
Methanol concentration	G	0.45
Sonication time	H	0.05
Particle size	I	0.08

With 6 degrees of freedom $t = 2.45$ giving:

$$= \text{effect} \pm \frac{2 \times 2.45 \times 0.081}{\sqrt{16}}$$

$$= \text{effect} \pm 0.10$$

Thus, three of the parameters have confidence intervals that do not include zero and are therefore significant. They are:-

Mobile phase composition (A)	-0.47 ± 0.10
Injection volume (C)	0.34 ± 0.10
Methanol concentration in sample preparation (G)	0.45 ± 0.10

7.8   Conclusions from the experiment

The standard deviation of the 16 determinations is 0.39. The residual SD is 0.081. The large difference between these values may be ascribed to the three significant effects. To bring down the SD of the determinations, the specification for these three parameters must be tightened up. A simple rule to achieve this is:

The effect estimate should not be more than the residual SD.

The residual SD is 0.081. To achieve an effect estimate of the same magnitude requires the following changes:

	Effect estimate	Present specification	New specification
Mobile phase composition	−0.47	±1%	$\pm \dfrac{.081}{.47} \times 1 = \pm 0.2\%$
Injection volume	0.34	±5 ul	$\pm \dfrac{.081}{.34} \times 5 = \pm 1.2 \mu l$
Methanol concentration	0.45	±2%	$\pm \dfrac{.081}{.45} \times 2 = \pm 0.4\%$

7.9 Did Serena achieve her aims?

Yes.

She has carried out only 16 trials but has investigated nine parameters, three of which were shown to have limits that needed tightening. She can now also widen the limits for the other parameters if it is desirable.

There was no evidence of outliers in the trial.

But . . .

There are no buts!

7.10 Which order should I use for the trials?

The negative answer to this question is to use the order in the design matrix – this order is highly correlated with the design matrix.

A better method is using random numbers. Randomize the order using the trial numbers, e.g., 1 to 16. With 16 trials obtain correlations between the trial number and all main effects. If the correlations are high, say above 0.3, randomize again and again until a suitable order is obtained. However, this is a lengthy process. The process is much easier and leads to usually lower correlations if we start by using one of the residual columns.

Let us see how the trial could have been ordered. We shall use the residual column J that is:

Design matrix order	1	2	3	4	5	6	7	8	9	10	11	12	13	14	15	16
Column J		−	+	+	−	−	+	−	−	−	+	+	+	+	−	+	−
Trial order	7	16	12	4	3	11	2	5	6	15	13	14	10	1	9	8

We note that the:

Minuses in column J correspond to trials 1 to 8

Pluses in column J correspond to trials 9 to 16

The trials pluses (or minuses) have then been ordered using random numbers. The trial order gives the following correlations between the variables and the trial order:

A	B	C	D	E	F	G	H	I
0.19	0.00	−0.11	−0.16	−0.03	−0.11	−0.11	0.03	0.19

These correlations are low so should there be no trend over time that could cause bias with the estimates. The time trend could also be estimated using multiple regression (see Chapter 9) and thus be removed from the residuals if it was significant, so that the residual standard deviation was not inflated.

7.11 How to obtain the designs

The designs are generated using a design vector that for the number of trials $(n) = 8, 12, 16, 20$ or 24 is given below.

$n = 8\ +\ +\ +\ -\ +\ -\ -$

$n = 12\ +\ +\ -\ +\ +\ +\ -\ -\ -\ +\ -$

$n = 16\ +\ +\ +\ +\ -\ +\ -\ +\ +\ -\ -\ +\ -\ -\ -$

$n = 20\ +\ +\ -\ -\ +\ +\ +\ +\ -\ +\ -\ +\ -\ -\ -\ -\ +\ +\ -$

$n = 24\ +\ +\ +\ +\ +\ -\ +\ -\ +\ +\ -\ -\ +\ +\ -\ -\ +\ -\ +\ -\ -\ -\ -$

To illustrate the operation of obtaining the design from the vector we shall use $n = 8$.

(a) Label seven columns A to G and seven rows 1 to 7.

(b) Write down the design vector under A, starting at row 1.

(c) Write down the design vector under B, starting at row 2. When row 7 is reached continue at row 1.

(d) Continue to use the vectors with columns C to G, also starting one row down from the previous column.

(e) Add a row of '$-$' in row 8.

Thus, the design is:

	A	B	C	D	E	F	G
1	+	-	-	+	-	+	+
2	+	+	-	-	+	-	+
3	+	+	+	-	-	+	-
4	-	+	+	+	-	-	+
5	+	-	+	+	+	-	-
6	-	+	-	+	+	+	-
7	-	-	+	-	+	+	+
8	-	-	-	-	-	-	-

7.12 Other uses of saturated designs

As well as being used in formulations on ruggedness experiments, saturated designs are also most useful in screening experiments.

Many investigations include major variables – those highly likely to be important, and minor variables, one or two of which might have an effect. Saturated designs are often used to screen a large number of minor variables. Any significant variables are then included in a more comprehensive design that could investigate interactions and curvature.

7.13   Problems

1. It is well known in the tobacco industry that moisture control is essential. Too low or too high a moisture level will make processing and packaging impossible. With this in mind, Dr Tabenda has been given the task of furthering his company's knowledge of the processing of the tobacco.

    Dr Tabenda consults the experts who point out that there is a multiplicity of variables that are important and wish him luck. However, during his discussions Dr Tabenda has classified the variables as major (ones that are very likely to affect processing) and minor (ones unlikely to affect processing). Of course, this classification is based upon their effect in the normal operating range. His next step is to carry out a screening experiment on the minor variables. There are six variables under consideration and he decides to use a Plackett–Burman design with 12 trials including two levels of each variable. The six variables together with their levels are:

A:	Top air temperature in drying oven:	$60\,°C$,	$70\,°C$
B:	Middle air temperature in drying oven:	$60\,°C$,	$70\,°C$
C:	Bottom air temperature in drying oven:	$60\,°C$,	$70\,°C$
D:	Baffle height:	$5\,cm$,	$10\,cm$
	(The baffle is designed to move tobacco from the edge into the middle of the drier.)		
E:	Baffle angle:	$20°$,	$30°$
F:	Drier	D_1,	D_2
	(Two nominally identical driers used in production)		

The responses (moisture values) in standard order were:

   23.0   25.8   25.2   23.5   27.1   24.8   22.0   24.6   28.4   28.3   26.4   25.3

Using the Plackett–Burman program:

   (a) Applying the 'residual columns' approach decide which effects are significant.
   (b) Look at the half-normal plot. All nonsignificant effects should form a straight line through zero. How many nonsignificant effects are there?
   (c) Include the appropriate number of nonsignificant effects in the calculation of the RSD. How does this change the width of the confidence interval for each effect?

2. Look at the effect of an outlier by decreasing the eighth value (28.4) to 22.4 and then repeating the analysis.

8

Regression analysis

8.1   Introduction

In this chapter we introduce the methods of regression and correlation, which relate one
variable to another. They are applied in many areas within science including calibration,
process investigation, microbiology, product stability and sensory analysis. They can be used,
for example, for the following purposes:

(i) to quantify the degree of association between two methods of assessment (e.g. sensory
and instrumental);

(ii) to obtain an equation to relate one variable to another (e.g. sensory quality and storage
time) and to predict one from the other;

(iii) to estimate the precision of a prediction.

(iv) to estimate the rate of decline of the sensory quality over time.

This chapter is a short exposition of regression analysis as a prelude to multiple regression.
A more detailed description is given in *Statistical Methods in Practice* by Boddy and Smith.

8.2   Example: keeping quality of sprouts

Farmers of the Wolds, a highly successful co-operative in the Midlands of England, sell their
vegetables through a nationwide chain of freezer centres. Having been hurt by some bad
publicity from customers who bought poor-quality Brussels sprouts, they suspect that the
freezer cabinets were not maintained at the correct temperature. They have commissioned a
shelf-life study from Bolton Laboratories.

Packets stored for various periods at different temperatures, representing well-maintained
and badly maintained freezers, are withdrawn, thawed and assessed by an expert panel for
colour, flavour (after cooking) and odour. The scoring system has a maximum score of 20,

Effective Experimentation: For Scientists and Technologists   Richard Boddy and Gordon Smith
© 2010 John Wiley & Sons, Ltd

which would be awarded to sprouts in perfectly fresh condition. Packets that are scored less than 12 are unacceptable.

We shall be trying to explain flavour scores, usually referred to as the response (y) in terms of weeks storage, usually referred to as independent variable (x).

The scores for flavour, along with the time of storage (in weeks) in a freezer cabinet at $-9\,^{\circ}$C, are given in Table 8.1.

Table 8.1 Flavour data for Brussels sprouts after frozen storage

Packet	Response (y) Weeks storage	Independent variable (x) Flavour score
1	2	15
2	2	19
3	4	17
4	4	16
5	7	13
6	7	15
7	9	11
8	12	14
9	12	11
10	15	12
11	20	7
12	26	6
Mean	10.0	13.0
SD	7.43	3.86

Before we can say much about a relationship between storage time and flavour we should inspect the data in the form of a scatter diagram with the regression line drawn on it as shown in Figure 8.1. The regression line is given by $y = a + bx$.

The regression line is:

$$\text{Flavour score} = 17.75 - 0.475 \times \text{weeks storage}$$

where 17.75 is the intercept and -0.475 is the slope.

8.3   How good a fit has the line to the data?

There are many criteria to determine the goodness of fit of the line:

Residual sum of squares and residual standard deviation

Correlation coefficient

Percentage fit

All of these show a mathematical relationship but let us first look at the residual sum of squares and residual standard deviation.

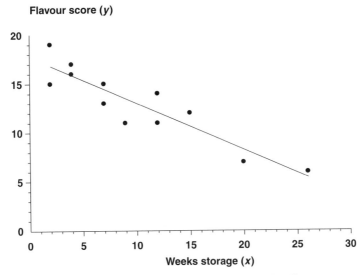

Figure 8.1 Scatter diagram and regression line.

8.4   Residuals

The criterion for determining the 'best' line is that it should be the closest to the points, the line with the lowest sum of squares of 'residuals'.

The 'residual' for each packet is the difference between the actual flavour score and the score predicted by the regression equation from the storage time, indicated by a vertical line from each point to the line in Figure 8.2.

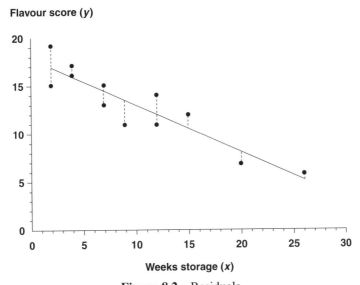

Figure 8.2   Residuals.

The residuals can also be calculated, as shown in Table 8.2.

Table 8.2 Calculation of residuals

Packet	Weeks (x)	Actual flavour (y)	Predicted flavour $y^* = 17.75 - 0.475x$	Residual $y - y^*$
1	2	15	16.80	−1.80
2	2	19	16.80	2.20
3	4	17	15.85	1.15
4	4	16	15.85	0.15
5	7	13	14.43	−1.43
6	7	15	14.43	0.57
7	9	11	13.48	−2.48
8	12	14	12.05	1.95
9	12	11	12.05	−1.05
10	15	12	10.62	1.38
11	20	7	8.25	−1.25
12	26	6	5.39	0.61

The residuals from the fitted line can be used in exactly the same way as deviations from a mean to determine a standard deviation. In regression analysis this is referred to as the residual standard deviation (RSD). The 'best' straight line minimizes the sum of the squared residuals or equivalently the RSD. Any other line would have a greater RSD. The line is often referred to as the 'least squares' line.

The residual standard deviation can be obtained directly from the residuals by squaring each residual and then computing the sum of the squared residuals, which is referred to as the residual sum of squares.

$$RSD = \sqrt{\frac{\text{Residual sum of squares}}{\text{Degrees of freedom}}}$$

The residual sum of squares $= (-1.80)^2 + 2.20^2 + 1.15^2 + \cdots + 0.61^2 = 26.69$

The degrees of freedom for the RSD are **two** less than the sample size; we 'lose' two degrees of freedom by estimating the values of 'a' and 'b' from the data.

$$\text{Degrees of freedom} = 12-2 = 10$$

$$RSD = \sqrt{\frac{26.69}{10}} = 1.63$$

We can interpret this in two ways:

(a) The 'average deviation' from the line is 1.63 score units.

(b) If data was collected for many packets, each with the same storage time, the scores would have a standard deviation of 1.63.

The value of 1.63 should be compared with the SD for the 12 packets, which was 3.86. Thus, fitting the regression line, thereby eliminating the effect of storage time on the flavour, has had a marked effect of reducing the original standard deviation of y.

8.5   Percentage fit

The percentage fit is a measure of how much of the original variability of y has been reduced by fitting the line.

$$\text{Original sum of squares} = (n-1) \times (\text{SD of flavour scores})^2$$
$$= 11 \times (3.86)^2$$
$$= 164.0$$

$$\text{Residual sum of squares} = 26.69$$

$$\text{Percentage fit} = \left(\frac{\text{Original sum of squares} - \text{Residual sum of squares}}{\text{Original sum of squares}} \right) \times 100$$
$$= \left(\frac{164.0 - 26.69}{164.0} \right) \times 100$$
$$= 84\%$$

8.6   Correlation coefficient

Correlation coefficients lie between -1 and $+1$. A value of 1.0 would indicate a perfect positive relationship, while a value of -1.0 would indicate a perfect negative relationship. Of course, we do not obtain perfect relationships because of random variation but we do require high correlations when we are trying to explain the values in one variable in terms of another variable. In this example the correlation coefficient is 0.915, which is high and one can safely conclude that storage time is affecting flavour although, later in the chapter, we shall confirm this using a significance test. Clearly, if the correlation coefficient had been near zero we could not have concluded there was a relationship.

One point should be emphasized strongly. The correlation coefficient is a measure of the linear association between two variables and is the same **no matter which of the variables is chosen as the response**. In fact, correlation is most useful when the variables cannot be categorized as either response or independent variables and also when a sample has been drawn from a population. However, where the levels of the independent variable can be deliberately varied by the researcher, the correlation coefficient is as much dependent upon the choice of levels as on the strength of relationship. Therefore, comparing correlation coefficients between different experiments is to be avoided.

One last point, in Chapter 9 we shall be looking at one response against many independent variables. Ideally, we would prefer that independent variables are truly independent and their correlations are zero.

8.7   Percentage fit – an easier method

An easier method of computing percentage fit is:

$$\text{Percentage fit} = 100r^2$$

where r is the correlation coefficient

$$= 100 \times (-0.915)^2$$
$$= 84\%$$

We shall see that percentage fit is an extremely useful measure in multiple regression as we seek to increase the percentage fit by adding further variables to the regression equation.

8.8   Is there a significant relationship between the variables?

In many situations the scientist would be uncertain whether there is a relationship or whether the value of the correlation coefficient or percentage fit is due to chance. A simple test for the association of two variables can be carried out using the percentage fit.

Null hypothesis -   Within the population of all packets there is no linear relationship between storage time and flavour score.

Alternative hypothesis -   Within the population there is a linear relationship between storage time and flavour score.

Test value -

$$= \sqrt{\frac{\text{Percentage fit} \times \text{Degrees of freedom}}{100 - \text{Percentage fit}}}$$

$$= \sqrt{\frac{84 \times 10}{100 - 84}}$$

$$= 7.24$$

Table values -   from Table A2 with 10 degrees of freedom (2 less than the sample size):

   2.23 at the 5% significance level;

   3.17 at the 1% significance level.

Decision -   Since the test value is higher than the table value at the 1% significance level we reject the null hypothesis at the 1% significance level.

Conclusion -   There is a relationship between storage time and flavour score.

8.9   Confidence intervals for the regression statistics

There are many confidence intervals that can be computed from regression statistics such as ones for the intercept, slope and future predictions. Usually, there is only one or two that are relevant to the situation. In this example the confidence interval for the slope is highly relevant.

$$\text{Confidence interval for the true slope} = b \pm (t)(\text{RSD})\sqrt{\frac{1}{(n-1)(\text{SD of } x)^2}}$$

t from Table A.2 with 10 degrees of freedom (the same as for the RSD)$=2.23$ at the 5% significance level

For a 95% confidence interval:

$$-0.475 \pm (2.23)(1.63)\sqrt{\frac{1}{11(7.435)^2}}$$

$$= -0.475 \pm 0.147$$

$$= -0.62 \text{ to } -0.33$$

Thus, there is a 95% chance that the flavour score decreases by between 0.33 and 0.62 units per week of storage. This interval is very wide, reflecting both the amount of variability about the line and the small sample size. The width of the confidence interval for the slope could be reduced by:

(i) increasing the sample size,
    –but the benefit is only in relation to $\sqrt{n}$.

(ii) increasing SD of x,
    –having a wider spread of storage times, provided it is still within a linear range.

(iii) reducing the residual standard deviation.

8.10   Assumptions

We now need to consider the assumptions behind regression analysis, a knowledge of which will help us to use regression analysis with safety. These assumptions are:

(a) The true relationship between the response and the independent variable is linear.

(b) The true value of the independent variable can be measured precisely whilst the measured value of the response is subject to sampling error. Regression analysis is not particularly sensitive to variation from this assumption especially with high correlations.

(c) The residuals are normally distributed.

(d) The residuals are independent of each other.

(e) The magnitude of the residuals is of the same order for all observations and does not increase with increasing magnitude of the independent variable.

It is unlikely that all, or indeed any, of these assumptions will be met precisely but it is essential that no one assumption is grossly violated.

Consider assumption (a): from Figure 8.1, the relationship seems to be reasonably linear. It is unlikely for example that true relationships will be precisely linear, but a linear relationship often provides a good fit to the data even if the true relationship is curved. This relationship might not, however, hold if we extrapolate outside the range of the data.

Assumption (b) is undoubtedly true since the storage time is measured exactly but the scores vary from packet to packet even after the same storage time.

The validity of assumption (c) can be assessed with either blob diagrams, histograms or a normal probability plot. The plot of residuals versus storage time is shown in Figure 8.3.

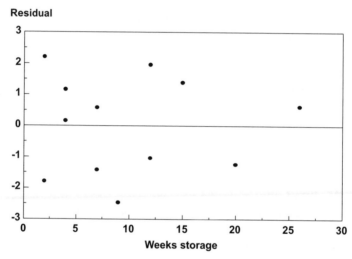

Figure 8.3   Plot of residuals against storage time.

We should not forget that an outlier will also render this assumption invalid and perhaps this kind of error is the most likely cause for violating the assumption. However, with this set of data further analysis reveals nothing untoward.

The validity of assumption (d) can be checked by plotting the residuals against the independent variable or, for that matter, any other variable that the researchers feel may affect the value of the response. Again there seems no cause for alarm.

Assumption (e) does not appear to be violated in this case.

We may therefore be reasonably assured that we have fitted the correct type of regression.

8.11   Problem

Yorkshire Spinners Ltd. buy triacetate polymer and convert it into different types of fabric. This process involves many stages, all of which affect the quality of the final product, but it has been shown that the critical stage is the first one in which polymer is heated, forced through minute orifices and wound onto bobbins as a fibre. The quality of the fibre may be represented by many parameters including birefringence, which is a measure of orientation of polymer molecules in the fibre and can be controlled by changing the speed of winding onto the bobbin – referred to as wind-up speed. The wind-up speed can be adjusted quickly but other associated adjustments – flow rate and temperature – cannot.

These adjustments, necessary to keep parameters other than birefringence constant, lead to a loss of production. It is necessary that the 'correct' adjustment be made to birefringence since over- or under-correction will lead to a further loss of production.

Making the 'correct' adjustment depends upon knowing accurately the relationship between birefringence and wind-up speed. This relationship is, however, dependent upon the type of machine.

A new machine is being developed by Yorkshire Spinners and the Research Manager has, absent-mindedly, independently asked four scientists – Addy, Bolam, Cooper and Dawson, to evaluate the relationship between birefringence and wind-up speed. He particularly wants an estimate of the slope that will be used in conjunction with hourly quality control checks on birefringence. If a change in birefringence is necessary the 'slope' will be used to indicate the required change in wind-up speed.

The designs are shown overleaf together with an analysis carried out by the company's dedicated but impractical statistician who cannot distinguish between important and irrelevant statistics. Carry out an evaluation of the designs.

Addy

Wind-up speed	150	160	170	180	190	200
Birefringence	70.1	75.3	77.0	80.6	87.2	89.8

Bolam

Wind-up speed	150	175	175	175	175	200
Birefringence	70.0	81.3	78.1	79.9	80.5	90.2

Cooper

Wind-up speed	150	150	150	200	200	200
Birefringence	70.4	70.8	69.0	91.5	88.9	89.4

Dawson

Wind-up speed	165	169	173	177	181	185
Birefringence	76.2	78.9	78.1	79.5	83.5	83.8

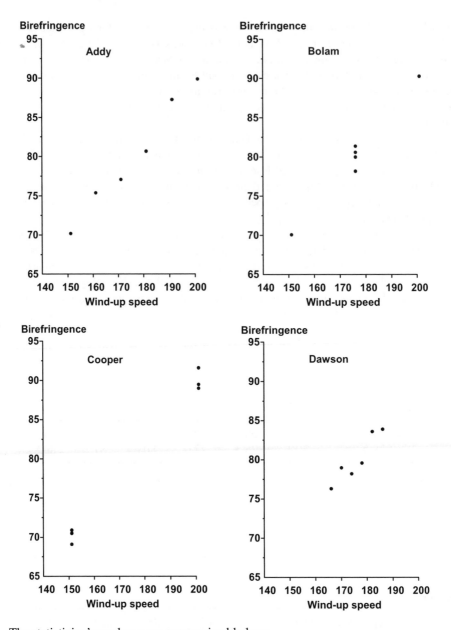

The statistician's analyses are summarized below:

	Addy	Bolam	Cooper	Dawson
Correlation coefficient	0.989	0.987	0.995	0.935
Intercept	11.1	9.3	10.5	13.5
Slope	0.394	0.404	0.397	0.380
Residual standard deviation	1.23	1.18	1.18	1.20
95% confidence interval for slope	±0.082	±0.093	±0.054	±0.199

9

Multiple regression: the first essentials

9.1 Introduction

In this chapter we are introduced to multiple regression, which must rank amongst the most powerful of statistical techniques. With multiple regression we seek to explain the values of the response variable in terms of many independent variables. It can therefore be used to analyse laboratory or plant experiments in which many variables have been deliberately varied. More importantly, it can be used to analyse plant data where variables have changed due to lack of control. This kind of data – often called 'dirty data' – is usually difficult to analyse since most of its features are outside the control of the researcher, but multiple regression can often be used to gain insight into the underlying process.

We shall use, as a case study, a badly designed experiment that bears little similarity to a well-designed experiment and gave results containing a number of features often associated with 'happenstance data'.

9.2 An experiment to improve the yield

Dr Ovi Dhose is the research manager for Pharma Fine Chemicals, a company that produces the active ingredient for pharmaceuticals. Their main contract is to produce Dioxy-A-123. Dr Dhose needs to increase the yield of the process in order to make the contract profitable.

The process consists of two reactions with the pH of the process determined after the primary reaction and acid added so that the secondary reaction is carried out under acidic conditions.

Dr Dhose has identified five variables that he believes are likely to affect the yield:

alkalinity of the primary reaction as measured by the pH;

weight of acid added to ensure that the secondary reaction is carried out under acidic conditions;

Effective Experimentation: For Scientists and Technologists   Richard Boddy and Gordon Smith
© 2010 John Wiley & Sons, Ltd

catalyst age: catalysts need regenerating after a certain time and Dr Dhose believes that they may be regenerating them at too long intervals;

temperature of the secondary reaction: Dr Dhose believes this needs optimizing;

reaction time in the secondary reaction: Dr Dhose believes this to be a key factor affecting yield.

Dr Dhose decides to run an experiment to determine the effect of these variables on the yield, with the aim of recommending conditions that would improve the yield from its present level of 92.0%. He considers that ten batches should be sufficient for his experiment.

He will start with a newly regenerated catalyst and use it until the end of the experiment. For each batch, he will obtain the pH at the end of the primary reaction and then decide on the weight of acid to be added, the reaction temperature and time of the secondary reaction.

At the end of the experiment he obtains the results shown in Table 9.1.

Table 9.1 Results from the experiment

				Reaction		
		Primary	Secondary	Secondary	Secondary	Secondary
Batch	Yield	pH	Weight of acid (kg)	Catalyst age	Temperature (°C)	Reaction time (min)
	A	B	C	D	E	F
1	91.5	9.8	1.6	1	45	70
2	92.9	9.2	1.3	2	40	70
3	97.3	8.8	1.1	3	50	70
4	94.1	9.0	1.2	4	45	50
5	89.9	9.8	1.6	5	40	50
6	94.8	9.0	1.2	6	45	50
7	93.7	9.6	1.4	7	50	50
8	92.2	10.0	1.7	8	50	60
9	90.9	9.6	1.5	9	45	60
10	91.7	9.2	1.4	10	40	60
Mean	92.90					
SD	2.16					

Note that the Yield values have been displayed in bold characters. The distinction between independent variables on the one hand and the response or dependent variable on the other hand is vitally important. When recommending a production strategy we must specify values for the independent variables and predict the level of response (i.e. yield) which can be expected if the strategy is adopted.

Dr Dhose's aims for this analysis are to:

(a) identify those variables that could be changed in order to increase the yield of the process;

(b) ascertain the nature of the relationship between these chosen variable(s) and the yield;

(c) recommend a strategy for the production of future batches.

9.3   Building a regression model

In simple regression where there is only one independent variable we seek the best equation involving that variable to enable us to predict values of the dependent variable.

In multiple regression we seek the best equation involving the independent variables together to enable us to predict the dependent variable. An equation that included all the independent variables might well give us a good prediction, but it might include variables that had no significant effect. In this chapter we shall seek the best equation that involves only those independent variables that are significant. The model will be built up by introducing one variable at each step. This approach is known as stepwise multiple regression.

9.4   Selecting the first independent variable

From our five independent variables we will first **select the one that appears to be most closely related to the dependent variable**. Before we perform any calculations let us examine the scatter diagrams in Figure 9.1, which illustrate these relationships. In each of the five

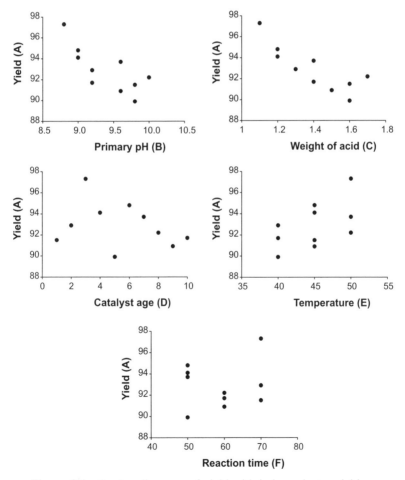

Figure 9.1   Scatter diagrams of yield with independent variables.

diagrams the dependent variable is on the vertical axis and one of the independent variables is on the horizontal axis. Each diagram illustrates the association between the yield of the process and one of the variables that might be having some influence upon it.

A visual inspection of the five diagrams reveals that:

(a) Those batches that had higher pH tend to have a low yields.

(b) Those batches that were manufactured with a high weight of acid tend to have low yields.

(c) No simple relationship is apparent between yield and age of the catalyst.

(d) A high yield may be associated with the use of a high secondary reaction temperature.

(e) Use of a higher reaction time may result in a higher yield, but the association between the two variables is not so strong and the visual impression is perhaps over-influenced by the batch that had a yield of 97.3 kg.

The drawing of scatter diagrams and the careful inspection of these diagrams can often bring to light features of a set of data that may otherwise go unnoticed and the utility of the simple scatter diagram should not be underestimated.

To proceed with multiple regression it is necessary to use a computer program. There are many packages that include routines for multiple regression, and it would be convenient if they all gave output in the same form, or even used the same one of several equivalent significance tests. Unfortunately, this is not the case. We shall illustrate the method using output from MultReg, a program written for Statistics for Industry, first using automatic stepwise variable selection.

The statistics at the first step are shown below.

Constant: 92.9			% fit: 0.0	d.f.: 9		Residual SD: 2.16
Variables in the equation				Variables available to add		
Variable	Coefficient	Decrease in % fit	Test value to delete	Variable	Increase in % fit	Test value to include
				B pH	57.61	3.30*
				C Weight	69.10	4.23**
				D Age	8.35	0.85
				E Temp	30.08	1.86
				F Time	1.52	0.35

In the output, the statistics at the top of the output relate to the data before any relationships are tested. The 'best fit' at that stage is the mean yield, 92.9. Nothing has been fitted so the % fit is 0.0. The residual standard deviation at that stage is the SD of yields, 2.16 with 9 degrees of freedom. (The mean and SD were shown in Table 9.1.)

The statistics associated with simple regression for each variable are shown. The regression of yield on weight would have a % fit of 69.10%, yield on age would have only 8.35%. So that we may determine whether any of these regressions are significant, a test value is given with an indication of its level of significance:

**	significant at the 1% level
*	significant at the 5% level
~	nearly significant (significant at the 10% level)

We should, however, be careful when using the word 'best' since this is defined by a statistical criterion and weight might not necessarily be the best variable to choose on scientific grounds. Many of the standard multiple regression programs automatically choose the 'best' variable, but some interactive programs allow the user to override this choice thereby enabling him/her to choose a variable that is considered to be more appropriate. We shall, however, initially use the standard procedure and choose weight (C), which has the highest per cent fit.

Before including weight we must first check on its significance using percentage fit and the test given in Chapter 8.

Percentage fit = 69.1	Degrees of freedom = 8
Test value = 4.23	Table value = 3.36 at the 1% significance level

Clearly, weight is significant and can be included in the model.

We note in passing that pH (B) has a test value of 3.30 that is also significant (at the 5% level). We must return to this later.

9.5 Relationship between yield and weight

Weight is entered into the equation and appears on the left side of the output. The resulting statistics are shown below.

Constant: 105.46			% fit: 69.10	d.f.: 8		Residual SD: 1.27
Variables in the equation				Variables available to add		
Variable	Coefficient	Decrease in % fit	Test value to delete	Variable	Increase in % fit	Test value to include
C Weight	−8.9722	69.10	4.23**	B pH	4.77	1.13
				D Age	0.61	0.38
				E Temp	24.31	5.08**
				F Time	1.52	0.60

Having established a significant relationship we can fit the simple regression line

$$\text{Yield} = 105.5 - 9.0\,\text{Weight}$$
$$\%\,\text{fit} = 69.1$$
$$\text{Residual SD} = 1.27$$

The residual SD is equivalent to a measure of the variability in yield levels at the same weight of added acid. By fitting an equation involving weight, the variability in yield has been reduced from 2.16, its original standard deviation, to its present value of 1.27. As we enter more variables, this value will decrease further. If we wish to make precise predictions for future batches we will need to make this value as small as possible.

The fitted equation is shown in Figure 9.2. We would examine the picture in some detail to see how well the line fits the data and whether there are any abnormalities in the residuals (the vertical distances of the points from the line).

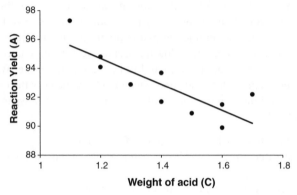

Figure 9.2 Regression of yield on weight of added acid.

In calculating a regression line we must remember that it is only an estimate of the true line and we would be well advised to examine certain confidence intervals to keep us in touch with reality. Clearly, in this example the slope of the regression line is important since it reflects the cost of reducing yield by increasing the weight of the added acid. In Chapter 8 we saw the formula for obtaining the 95% confidence interval for the slope.

This is:

$$= -9.0 \pm 4.9$$
$$= -13.9 \text{ to } -4.1 \text{ kg/kg.}$$

Thus, each kg of acid added could reduce the yield by between 4.1 and 13.9 kg. Another feature of great interest is the residuals and these are shown in Table 9.2.

Table 9.2 Table of residuals after fitting the equation. Yield $= 84.9 - 9.0$ weight

Batch	Weight	Actual yield	Predicted yield	Residual
1	1.6	91.5	91.11	0.39
2	1.3	92.9	93.80	−0.90
3	1.1	97.3	95.59	1.71
4	1.2	94.1	94.69	−0.59
5	1.6	89.9	94.11	−1.21
6	1.2	94.8	94.69	0.11
7	1.4	93.7	92.90	0.80
8	1.7	92.2	90.21	1.99
9	1.5	90.9	92.00	−1.10
10	1.4	91.7	92.90	−1.20

9.6   Model building

A regression equation such as the one pictured in Figure 9.2 is often called a model (or mathematical model) because it is a representation in algebraic form of what happens on the plant. Since clearly no simple equation can be an exact model of the plant; the art of multiple regression analysis may be thought of as model building with the objective of finding as simple a model as possible, which gives an adequate representation of the behaviour of the plant. The remainder of this chapter is concerned with a step-by-step method of investigating whether, starting from the best simple model we have currently found, we can find a slightly more complicated model that will give a statistically significantly better fit to the experimental data.

We continue by looking to see if the inclusion of a second independent variable in the model would give a worthwhile improvement in fit to the yield data.

9.7   Selecting the second independent variable

At this juncture we might be tempted to select pH since it had the second highest test value at the first step. (The previous output, however, suggested that pH would not be successful.) If we follow this path we obtain the following output:

Constant: 72.46				% fit: 73.87	d.f.: 7		Residual SD: 1.25
Variables in the equation					Variables available to add		
Variable	Coefficient	Decrease in % fit	Test value to delete		Variable	Increase in % fit	Test value to include
B pH	5.0000	4.77	1.13		D Age	0.07	0.13
C Weight	−18.9722	16.26	2.09∼		E Temp	20.21	4.53**
					F Time	3.58	0.98

The equation becomes:

$$Yield = 72.50 - 19.0\,Weight + 5.0\,pH$$
$$\% \text{ fit} = 73.9$$
$$Residual\ SD = 1.25$$

Now this is quite alarming! The coefficient of weight has altered from −9.0 to −19.0, which indicates that the model is unstable. Even worse, the coefficient of pH is positive, while the relationship on the scatter diagram in Figure 9.1 is negative.

To add to our woe, all we have gained is a very small increase in per cent fit and little change in the residual SD. Clearly it is unwise to include both weight and pH – a point we will return to later.

In considering which second variable to include we should not be looking at the relationships of the candidates with yield, but how best they explain the variability remaining in yield after fitting the equation involving weight.

Let us look at the residuals and see how these relate to the remaining variables. They are plotted in Figure 9.3.

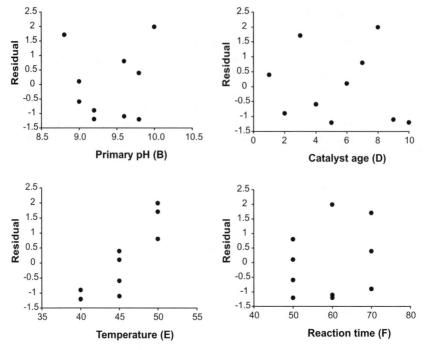

Figure 9.3 Residuals after fitting weight, plotted against the other independent variables.

Clearly if we are to improve the fit we must reduce the magnitude of the residuals so it is the scatter diagrams in Figure 9.3 that are now important, not the original ones in Figure 9.1. Figure 9.3 shows that one variable (temperature) has a strong relationship with the residuals and this was borne out by the program. Let us examine the output (after the entry of weight) again.

Constant: 105.46			% fit: 69.10	d.f.: 8		Residual SD: 1.27
Variables in the equation				Variables available to add		
Variable	Coefficient	Decrease in % fit	Test value to delete	Variable	Increase in % fit	Test value to include
C Weight	−8.9722	69.10	4.23**	B pH	4.77	1.13
				D Age	0.61	0.38
				E Temp	24.31	5.08**
				F Time	1.52	0.60

We can see how the automatic selection procedure chooses the next variable to be entered using per cent fit and a significance test.

We see that temperature is the best variable, with an increase in % fit of 24.31 and a test value of 5.08. Let us now see whether this is significant by comparing it with a value from Table A.2 with 7 degrees of freedom. The table value is 2.36 at a 5% significance level. There is no doubt about the significance of temperature and there is also no doubt about the other variables being not significant.

Temperature is therefore entered into the equation.

Constant: 93.19			% fit: 93.41	d.f.: 7		Residual SD: 0.63
Variables in the equation				Variables available to add		
Variable	Coefficient	Decrease in % fit	Test value to delete	Variable	Increase in % fit	Test value to include
C Weight	−8.6093	63.33	8.20**	B pH	0.67	0.83
E Temp	0.2613	24.31	5.08**	D Age	1.22	1.16
				F Time	1.52	1.34

The new regression equation is

$$\text{Yield} = 93.2 - 8.61\,\text{Weight} + 0.261\,\text{Temp}$$
$$\text{\% fit} = 93.4$$
$$\text{Residual SD} = 0.63$$

The significance of temperature implies that the simple model predicting yield from the weight of added acid alone is an inadequate model. We are now able to conclude that **both** the variation in secondary reaction temperature **and** the variation in weight of added acid are contributing to the batch-to-batch variation in yield.

We can now proceed to search for a third variable to improve the fit still further.

Time has the highest per cent fit but its test value of 1.34 is somewhat below the table value of 2.45 obtained from Table A.2 with 6 degrees of freedom at the 5% significance level.

The stepwise procedure now terminates.

We notice that the coefficient for weight of −8.61 is little changed from the simple equation in which it had a coefficient of −9.0. This is a good sign, showing that the final equation is stable and that it is worthwhile to examine the residuals as shown in Table 9.3.

Table 9.3 Residuals from the final equation

Batch	Weight of added acid	Secondary reaction temperature	Actual yield	Residual	Predicted value (with 95% confidence interval for true mean yield)
1	1.6	45	91.5	0.32	91.2 ± 0.7
2	1.3	40	92.9	0.45	92.5 ± 0.8
3	1.1	50	97.3	0.51	96.8 ± 1.0
4	1.2	45	94.1	−0.52	94.6 ± 0.7
5	1.6	40	89.9	0.03	89.9 ± 0.9
6	1.2	45	94.8	0.18	94.6 ± 0.7
7	1.4	50	93.7	−0.51	94.2 ± 0.8
8	1.7	50	92.2	0.58	91.6 ± 1.1
9	1.5	45	90.9	−1.14	92.0 ± 0.5
10	1.4	40	91.7	0.11	91.6 ± 0.8

Since this will turn out to be the final equation we have included 95% confidence intervals for the true mean yield.

We notice that batches 3, 4, 6 and 7 have confidence intervals in which the lower limit exceeds the present yield mean of 92.0%. This gives Dr Dhose hope that the experiment will give a successful conclusion.

The confidence intervals assume that we have the correct form of the equation. This assumption could be invalid in the following situations:

(i) One or more important variables have been left out of the equation because we failed to recognize their importance and include them in the list of independent variables.

(ii) The regression equation of yield against weight or temperature is not linear but has some curved relationship. Curvature is dealt with in Chapter 10.

(iii) There may be interactions present that we have failed to consider in our analysis.

(iv) The effect of, say, weight or temperature on yield is dependent upon the level of a third variable. In this example the weight of added acid in the secondary reaction was dependent upon the pH in the primary reaction. We must be careful about our conclusions.

9.8   An alternative model

Let us now return to our analysis and look for alternative models. We saw, when we looked for the first variable to enter in the model, that pH was also significant and had a per cent fit that was almost as high as that for weight.

We shall therefore ignore weight and choose pH to enter in the model.

Constant: 130.38			% fit: 57.61	d.f.: 8		Residual SD: 1.49
Variables in the equation				Variables available to add		
Variable	Coefficient	Decrease in % fit	Test value to delete	Variable	Increase in % fit	Test value to include
B pH	−3.9868	57.61	3.30*	C Weight	16.26	2.09~
				D Age	2.04	0.59
				E Temp	36.00	6.28**
				F Time	0.59	0.31

This gives us the following equation:

$$\text{Yield} = 130.4 - 3.99\,\text{pH}$$
$$\text{\% fit} = 57.6$$
$$\text{Residual SD} = 1.49$$

Entry of pH into the model results in weight being no longer significant. However, temperature is again the best variable, and is entered into the equation.

Constant: 118.03		% fit: 93.61	d.f.: 7	Residual SD: 0.62	
Variables in the equation			Variables available to add		
Variable	Coefficient	Decrease in % fit	Test value to delete	Variable	Increase in % fit	Test value to include
B pH	−4.1960	63.53	8.34**	C Weight	0.47	0.69
E Temp	0.3180	36.00	6.28**	D Age	2.64	2.05~
				F Time	0.55	0.75

The new equation is:

$$Yield = 118.0 - 4.20\,pH + 0.318\,Temp$$
$$\%\ fit = 93.6$$
$$Residual\ SD = 0.62$$

Catalyst age fails to be significant at the 5% level but has a fairly high test value and we should keep a watchful eye on this variable. The coefficient for catalyst age is negative, thus indicating the possibility that yield decreases with age of catalyst.

9.9   Limitations to the analysis

Let us remind ourselves of the two possible equations that have been obtained.

$$Yield = 93.2 - 8.61\,Weight + 0.261\,Temp$$
$$Yield = 118.0 - 4.20\,pH + 0.318\,Temp$$

Now both sets of equations and advice cannot be correct. In fact it is likely that, at the most, only one equation is reasonable. This conflicting advice is due to **bad design** as a result of operational considerations mentioned on page 1. This is apparent from examining Figure 9.4.

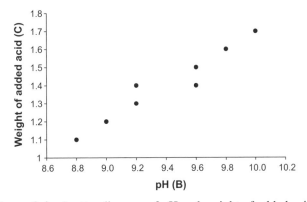

Figure 9.4   Scatter diagram of pH and weight of added acid.

In Figure 9.4 we see that every time that weight is changed so is pH in the same way. Thus, it is impossible to say which of the variables is affecting yield. The correlation between weight and pH is 0.97, which is very high. This could have been avoided by Dr Dhose since both variables are under his control, but he has failed to do so and we are now becoming aware of the shortcomings of his design. These shortcomings can be seen by examining the intercorrelation matrix. This is a matrix of correlations between independent variables as shown in Table 9.4.

Table 9.4 Intercorrelation matrix

	B	C	D	E	F
B pH	*	0.97	0.20	0.07	−0.06
C Weight		*	0.26	−0.07	0.00
D Age			*	0.04	−0.44
E Temp				*	0.00
F Time					*

The ideal intercorrelation is zero. All of our correlations are small except for those between weight and pH (0.97) that we know has given us problems and time and age (−0.44), which might have given problems. What value of intercorrelation will cause difficulties? Unfortunately, there is no easy answer to this question and the only way of determining this is to run the regression program several times searching for alternative models that are also partially feasible.

Perhaps this section could be summarized by:

high correlations between dependent and independent variables are a state of nature and are good;

high correlations between independent variables are caused by the experimenter and are bad.

Since a well-designed experiment has zero or small intercorrelations between the independent variables it is good practice always to determine and examine the intercorrelation matrix **before** carrying out the experiment.

9.10   Was the experiment successful?

Yes.

It has been shown that yields can be obtained well above the present value of 92.0% providing relevant experimental conditions are chosen.

Higher reaction temperature will increase the yield.
Either pH or weight of acid changes yield.
Reaction time is not an important variable.

But ...

We do not know whether pH or weight is important due to a bad design. We will explore this further in Chapter 16.

We need more information (more trials) to determine whether catalyst age is an important variable. Clearly, Dr Dhose needs to ascertain this to decide when to change the catalyst.

We did not consider curvature or interaction in the design or analysis.

9.11   Problems

1. Kandie Confectionery sell a special confectionery on a nationwide basis. They are concerned that sales vary from region to region and have recently appointed a committee to investigate the reasons for this disparity. The committee suggested several possible causes:

the size of the region (its population);
the number of salespersons employed;
the wealth of the region;
the quality (as represented by lack of complaints);
the number of competitors;
the amount spent on advertising.
It was decided to collate data relating to these variables and this is given below.

Region A	Population (m) B	Salespersons C	GDP (£bn) D	Complaints E	Competitors F	Advertising (£m) G	Sales (£m) H
1	1.2	5	17.2	29	3	4.0	147.3
2	5.6	34	71.1	134	3	16.0	612.3
3	0.9	7	13.9	12	5	2.0	52.5
4	3.4	15	58.1	109	4	14.5	356.4
5	2.7	16	41.3	86	4	6.0	208.4
6	7.2	54	115.2	173	5	10.0	371.9
7	4.4	20	55.0	70	2	15.5	646.3
8	1.4	8	19.9	34	3	6.0	195.2
9	2.9	22	31.6	46	2	8.0	388.9
10	4.1	18	48.8	164	5	14.5	269.0
11	1.7	10	21.8	41	3	7.0	230.1
12	2.6	20	42.1	21	1	5.5	366.3

Using stepwise multiple regression find all the alternative multiple regression equations. Consider their value in showing which of the independent variables affect sales.

2. In Problem 1 did you spot what was causing the ambiguity? Several of the variables were a consequence of the size of the region as measured by its population. To eliminate this dependence, we need to standardize the data by dividing by the population to give sales per head, advertising per head, etc. Only two variables, population and number of competitors, remained unstandardized. The data after standardizing is:

Region A	Population (m) B	Salespersons per m. C	GDP per head (£'000) D	Complaints per m. E	Competitors F	Advertising per head (£) G	Sales per head (£) H
1	1.2	4.2	14.3	24.2	3	3.33	122.8
2	5.6	6.1	12.7	23.9	3	2.86	109.3
3	0.9	7.8	15.4	13.3	5	2.22	58.3
4	3.4	4.4	17.1	32.1	4	4.26	104.8
5	2.7	5.9	15.3	31.9	4	2.22	77.2
6	7.2	7.5	16.0	40.0	5	1.39	51.7
7	4.4	4.5	12.5	15.9	2	3.52	146.9
8	1.4	5.7	14.2	24.3	3	4.29	139.4
9	2.9	7.6	10.9	15.9	2	2.76	134.1
10	4.1	4.4	11.9	40.0	5	3.54	65.6
11	1.7	5.9	12.8	24.1	3	4.12	135.4
12	2.6	7.7	16.2	8.1	1	2.12	140.9

Carry out a stepwise multiple regression analysis on the data. How are sales affected by the standardized independent variables? How many different models do you reach this time?

10

Designs to generate response surfaces

10.1 Introduction

Two-level factorial designs, which we met in Chapters 3–6, are extremely efficient in separating the important variables from the trivial ones but do not indicate how the response changes over the experimental space nor do they enable us to determine what conditions will give the best response. In order to achieve this, we need to extend the designs to three or four levels, and then amend our analysis in order to obtain a response surface – a surface showing how the change in level of variables alters the response. In this chapter we look at a mixture of two- and three-level designs as well as the method of achieving response surfaces using multiple regression analysis followed by contour diagrams.

10.2 An example: easing the digestion

A pharmaceutical company, Pharmamart, has asked one of their development scientists, Dev Lauper, to produce a formulation for a new indigestion pill. In particular they are concerned with two responses:

The **crushing strength** of the tablet has to meet specifications. Too low a level, below 8 kiloPascals (kPa), might result in the tablet crumbling when packed, so they are seeking a high crushing strength.

The **dissolution time** of the tablet. This is the time to dissolve a tablet under specific test conditions and simulates, to some degree, how the tablet would behave in the body. A dissolution time of between 20 and 30 min is specified.

Dev has already carried out some preliminary work and has established three variables that he believes to be important. These are:

Effective Experimentation: For Scientists and Technologists   Richard Boddy and Gordon Smith
© 2010 John Wiley & Sons, Ltd

Bulking agent (A): There is a choice of lactose or cellulose.

Magnesium stearate (B): This could vary from 0.5% to 1%.

Hydrogenated vegetable oil (C): This could range from 1% to 4%.

The aims of the investigation are therefore to establish which of these variables have an effect on the two responses, and in what way, and what conditions will give the best tablet.

In choosing the levels to use in the experiment, Dev considers the following:

He has a choice of two bulking agents, so can have only two levels of that variable.

The relationship between the variables and the responses may be nonlinear and interactive. He will need more than two levels of the other variables.

He decides to carry out an experiment using a traditional design with two levels of A (lactose and cellulose), three levels of B (0.5, 0.75 and 1.0%) and three levels of C (1.0, 2.5 and 4.0%). Notice that the levels of the quantitative variables will be equally spaced. The design is shown diagrammatically in Figure 10.1.

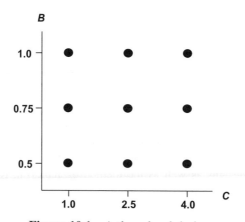

Figure 10.1   A three-level design.

These 9 conditions are carried out with both lactose and cellulose, making 18 trials in total.

These, together with the responses, are summarized in Table 10.1. The bulking agent is, of course, lactose or cellulose. When we come to analyse the results using multiple regression, which we met in Chapter 9, we will need those choices to be given numerical values. For convenience we have given a value of 1 for lactose and 2 for cellulose.

There are a number of methods of analysing the results of this experiment but by far the simplest is multiple regression. For each response we will try to fit the terms:

$$\begin{array}{ll} \text{Main effects:} & A, B, C \\ \text{Interactions:} & AB, AC, BC \\ \text{Quadratics:} & B^2, C^2 \end{array}$$

There is, of course, no quadratic for variable A since it has only 2 levels.

The intercorrelation matrix for these terms is shown in Table 10.2.

The intercorrelations are all zero, the perfect value. This is not too surprising; well-designed experiments should give zero or very low inter-correlations.

Table 10.1 Experimental conditions and responses

Bulking agent (A)	Magnesium stearate (B)	Hydrogenated vegetable oil (C)	Crushing strength	Dissolution time
1	0.50	1.0	13.47	30.6
1	0.50	2.5	10.04	27.0
1	0.50	4.0	8.45	20.5
1	0.75	1.0	14.02	34.5
1	0.75	2.5	9.23	33.6
1	0.75	4.0	12.26	28.7
1	1.00	1.0	3.66	34.1
1	1.00	2.5	8.84	37.1
1	1.00	4.0	14.75	38.3
2	0.50	1.0	12.18	29.0
2	0.50	2.5	9.20	27.5
2	0.50	4.0	8.06	17.7
2	0.75	1.0	12.21	31.6
2	0.75	2.5	8.78	27.4
2	0.75	4.0	12.95	20.3
2	1.00	1.0	5.24	30.4
2	1.00	2.5	7.16	32.7
2	1.00	4.0	14.12	19.2

The sharp-eyed reader will be surprised that the correlation between B and B^2 is zero when clearly as B increases so does B^2. The quadratic term, denoted by B^2, is actually

$$(B-\bar{B})^2$$

This version is used so that it represents the additional quadratic component after fitting a linear relationship and is independent of B.

Similarly the interaction terms are of the form

$$(B-\bar{B})(C-\bar{C})$$

Table 10.2 Intercorrelation matrix

	A	B	C	B^2	C^2	AB	AC	BC
A	1.0	0.0	0.0	0.0	0.0	0.0	0.0	0.0
B		1.0	0.0	0.0	0.0	0.0	0.0	0.0
C			1.0	0.0	0.0	0.0	0.0	0.0
B^2				1.0	0.0	0.0	0.0	0.0
C^2					1.0	0.0	0.0	0.0
AB						1.0	0.0	0.0
AC							1.0	0.0
BC								1.0

10.3 Analysis of crushing strength

We have previously used multiple regression for investigations in which several variables have been explored to determine which ones had an effect on the response. In response surface fitting we assume all the variables are significant, unless there is evidence to the contrary, and try to obtain the best response surface using **all** interactions and quadratics. It is therefore generally good practice to start with all the terms entered into the model.

Let us look at this procedure using crushing strength as the response and using the same computer program, *MultReg*, as in Chapter 9.

The first step shows the following:

Constant: 19.53	% fit: 89.57		d.f.: 9		Residual SD: 1.41	
Variables in the equation				**Variables available to add**		
Variable	Coefficient	Decrease in % fit	Test value to delete	Variable	Increase in % fit	Test value to include
A	−1.7611	0.75	0.81			
B	18.8683	2.83	1.56			
C	−11.5364	4.69	2.01~			
B^2	−31.6400	9.14	2.81*			
C^2	0.9211	10.04	2.94*			
AB	1.1933	0.16	0.37			
AC	0.1322	0.07	0.24			
BC	9.7033	61.89	7.31**			

All the variables, including quadratics and interactions, have been entered into the model. In Chapter 9, as we did not know which variables would be significant, we built up the model from the first variable, deciding which variable would significantly improve the model by being included. This time we inspect the list of variables and identify those that, if removed from the model, would significantly weaken the model.

The table value against which each test value should be compared is taken from Table A.2 with the same degrees of freedom as the residual degrees of freedom.

With 9 degrees of freedom, the table value from Table A.2 is 2.26 at the 5% significance level and 3.25 at the 1% level.

With low intercorrelations the importance of any term can be judged by the size of the test value. Clearly this response surface is going to be dominated by the interaction BC, while terms such as AB are going to have negligible effects.

We also notice that the interaction BC and the quadratic terms B^2 and C^2 are significant. We will also therefore include B and C to complete the model.

When any interaction or quadratic is significant, then the main effects of the variable(s) should also be kept in the equation.

However, there is no indication of A having an effect on the response since all the terms that include A have low test values.

So, despite the strong prior beliefs about the importance of bulking agent (A) we will delete it and all its interactions from the equation. This gives:

Constant: 16.89	% fit: 88.59		d.f.: 12	Residual SD: 1.28		
Variables in the equation				Variables available to add		
Variable	Coefficient	Decrease in % fit	Test value to delete	Variable	Increase in % fit	Test value to include
B	20.6583	2.83	1.73	A	0.75	0.88
C	−11.3381	4.69	2.22*	AB	0.16	0.39
B^2	−31.6400	9.14	3.10*	AC	0.07	0.26
C^2	0.9211	10.04	3.25**			
BC	9.7033	61.89	8.07**			

This confirms our conclusions. There are, however, some interesting comparisons that can be made between this and the previous analysis.

(a) The per cent fit for each effect is unchanged. This is because all the intercorrelations are zero.

(b) The test values have changed slightly. This is because of the change of the residual standard deviation with fewer terms in the equation.

(c) The significance of some terms changes due to the change in the test value and the degrees of freedom. However, significance testing is only a guide in response surface fitting and such changes make no changes to the conclusions.

We shall use the equation including terms for B and C only. This is:

$$y = 16.9 + 20.7B - 11.3C - 31.6B^2 + 0.92C^2 + 9.70BC$$

The best way of representing the equation is a response surface or contour diagram as shown in Figure 10.2.

The response surface is similar to a contour map of a geographical region in which the contours (lines) indicate the height above sea level. In our diagram points on the same contours have the same crushing strengths. We are looking for high crushing strength and these occur with low magnesium stearate and low vegetable oil or with both high.

Another way of showing a response surface is as a 3-dimensional figure (Figure 10.3). It shows the shape of the surface in a more graphic way, but is little help in recognizing where the optimum regions of the experimental region occur.

However, the criterion is that the crushing strength is above 8.0 and we have fitted a model without looking at confidence intervals or checking the residuals. These are given in Table 10.3.

There is one residual that is worthy of note – that for observation 4. It might be well worthwhile to remove this observation, carry out the multiple regression again and then

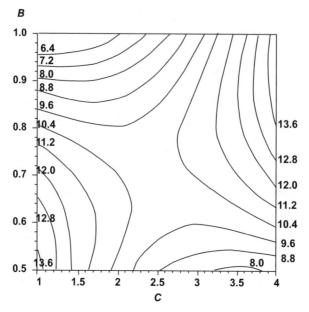

Figure 10.2   Response surface for crushing strength.

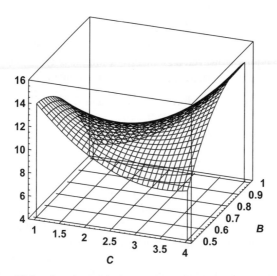

Figure 10.3   An alternative presentation of a response surface.

compare the contour diagram with the original. If the response surface has changed consider-ably it means that this observation is having an undue influence.

We notice that the confidence interval is either ±1.47 or ±1.76. For ease of use we shall use ±1.6. The specification referred to a crushing strength greater than 8.0 kPa so for safety we should consider only conditions which give a predicted strength in excess of 9.6 kPa.

Table 10.3 Predicted values and residuals

Trial	Observed value	Predicted value	95% conf. interval for the mean (±)	Residual
1	13.47	13.75	1.76	−0.28
2	10.04	8.85	1.47	1.19
3	8.45	8.10	1.76	0.35
4	14.02	11.45	1.47	2.57
5	9.23	10.19	1.47	−0.96
6	12.26	13.08	1.47	−0.82
7	3.66	5.20	1.76	−1.54
8	8.84	7.58	1.47	1.26
9	14.75	14.11	1.76	0.64
10	12.18	13.75	1.76	−1.57
11	9.20	8.85	1.47	0.35
12	8.06	8.10	1.76	−0.04
13	12.21	11.45	1.47	0.76
14	8.78	10.19	1.47	−1.41
15	12.95	13.08	1.47	−0.13
16	5.24	5.20	1.76	0.04
17	7.16	7.58	1.47	−0.42
18	14.12	14.11	1.76	0.01

10.4 Analysis of dissolution time

We again use multiple regression analysis and initially fit all the terms. The result is as follows:

Constant: 3.86	% fit: 91.80	d.f.: 9		Residual SD: 2.40	
Variables in the equation				**Variables available to add**	
Variable	Coefficient	Decrease in % fit	Test value to delete	Variable	Increase in % fit	Test value to include
A	12.3889	20.67	4.76**			
B	40.6667	20.48	4.74**			
C	4.1667	27.17	5.46**			
B^2	−10.8000	0.29	0.56			
C^2	−1.3222	5.58	2.47*			
AB	−15.5333	7.13	2.80*			
AC	−2.4556	6.41	2.65*			
BC	4.8000	4.08	2.12~			

In this case our decision is easy: we should include all the terms in the equation to give:

$$y = 3.86 + 12.4A + 40.7B + 4.17C - 10.8B^2 - 1.32C^2 - 15.5AB - 2.46AC + 4.80BC$$

The confidence intervals and residuals obtained using this equation are given in Table 10.4.
The residuals show no discernible pattern. The width of the confidence interval is, however, alarming, being approximately ±4 min, while the specification is 20 to 30 min.

Table 10.4 Predicted values and residuals

Trial	Observed value	Predicted value	95% conf. interval for the mean ($\pm$)	Residual
1	30.6	28.90	4.30	1.70
2	27.0	28.12	3.51	−1.12
3	20.5	21.40	4.30	−0.90
4	34.5	33.01	3.51	1.49
5	33.6	34.03	3.14	−0.43
6	28.7	29.11	3.51	−0.41
7	34.1	35.77	4.30	−1.67
8	37.1	38.59	3.51	−1.49
9	38.3	35.47	4.30	2.83
10	29.0	31.07	4.30	−2.07
11	27.5	26.61	3.51	0.89
12	17.7	16.20	4.30	1.50
13	31.6	31.29	3.51	0.31
14	27.4	28.63	3.14	−1.23
15	20.3	20.03	3.51	0.27
16	30.4	30.17	4.30	0.23
17	32.7	29.31	3.51	3.39
18	19.2	22.50	4.30	−3.30

Our response surface includes three variables. We must now represent this by using a series of contour diagrams. Since bulking agent is a qualitative variable we shall produce a contour diagram for each agent. For lactose we fix the value of A at 1. This gives the modified equation:

$$y = 16.3 + 25.2B + 1.71C - 10.8B^2 - 1.32C^2 + 4.80BC$$

This gives a contour diagram shown in Figure 10.4.

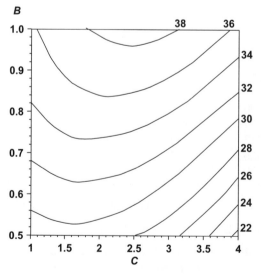

Figure 10.4   Contour diagram of dissolution time for lactose.

There is only one corner that will give reasonable certainty of meeting the specification. However, there are no conditions in this region that will satisfy the specification for crushing strength. To produce the contour diagram for cellulose we fix the value of A in the equation at 2 to give:

$$y = 28.7 + 9.7B - 0.75C - 10.8B^2 - 1.32C^2 + 4.80BC$$

This gives a contour diagram shown in Figure 10.5.

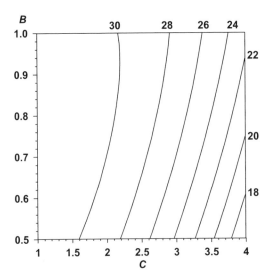

Figure 10.5   Contour diagram of dissolution time for cellulose.

The region between dissolution times of 24 and 26 min will be suitable providing that high values are chosen for both B and C to give high crushing strength. Thus, our choice of conditions is:

Bulking agent (A):	Cellulose
Magnesium stearate (B):	1.0%
Hydrogenated vegetable oil (C):	3.6%

These conditions will produce tablets with predicted crushing strength of 12.0 kPa and predicted dissolution time of 25.0 min with 95% confidence intervals of approximately ± 1.7 kPa and ± 4.2 min, respectively.

10.5   How many levels of a variable should we use in a design?

In the previous experiment we used three levels for magnesium stearate and hydrogenated vegetable oil. The choice of three levels for these two variables was correct. The variables were important, the relationship was curved and all the points were relevant in estimating the curvature and predicting the optimum. Having more than three levels would have served no useful purpose in this experiment.

In what circumstances should we use four levels? This will be necessary when it is believed that either the high or low level will give a response that is so far removed from the optimum that it may have to be ignored.

In what circumstances should only two levels be included? This will often be done in the initial stages of an investigation when we have no prior knowledge about the importance of the variables. Of course, we also use only two levels when the variable is **qualitative** with only two categories.

10.6   Was the experiment successful?

Yes.

Dev designed an experiment that covered the experimental space for the two quantitative variables at both levels of the bulking agent.

He was able to determine which variables were significant (B and C including their quadratics and their interaction for crushing strength; A, B and C including interactions and quadratics for dissolution time).

From the equations and contour diagrams he could identify the conditions for highest crushing strength and those that met the specification for dissolution time, and hence the recommended conditions to satisfy both criteria.

But...

He could not be certain of the exact conditions, so had to take account of the confidence intervals of prediction.

The optimum conditions were at or near an extreme value of magnesium stearate, so the uncertainty would be greater than if they had been closer to the centre of the experimental space.

In other words, his choice of conditions for magnesium stearate was only just acceptable. He would have been better using 0.75 to 1.25 instead of 0.5 to 1.0. However, this might just be wise using hindsight.

10.7   Problem

In the manufacture of Rudyard's Rock Cakes, three ingredients are believed to have an effect on the acceptability to consumers. These are:

A: The amount of sugar;
B: The amount of spice;
C: The amount of a secret ingredient called ATN.

Eighteen batches were prepared using combinations of three levels of the ingredients. Samples of each batch were assessed by a panel of five employees who gave scores up to 10 for acceptability and for 'crumbliness'.

The aim of the product development department is to produce cakes with as high an acceptability score as possible while keeping crumbliness scores within a specification of between 4 and 7. (A score below 4 indicates that cakes do not easily come apart when bitten; above 7 they have fallen apart before reaching the mouth.)

The panel mean scores were as follows:

Batch	Sugar (A)	Spice (B)	ATN (C)	Acceptability (D)	Crumbliness (E)
1	2	0.8	10	3.0	6.2
2	2	1.2	20	6.0	5.8
3	2	1.6	15	6.6	4.8
4	2	0.8	15	3.6	5.6
5	4	1.2	20	8.4	3.2
6	4	1.6	20	7.6	5.0
7	4	0.8	10	6.2	5.6
8	4	1.2	10	8.2	7.4
9	6	1.6	15	8.2	5.6
10	6	0.8	15	8.4	4.6
11	6	1.2	20	9.2	4.6
12	6	1.6	20	9.2	3.6
13	4	1.2	15	8.6	6.0
14	6	1.6	10	8.6	6.4
15	2	1.6	10	6.6	6.8
16	4	0.8	20	7.0	2.6
17	6	1.2	10	8.4	7.2
18	4	1.6	15	7.8	5.6

(a) By using multiple regression (with all quadratics and interactions) determine which variables have a significant effect on acceptability.

(b) Obtain a contour plot and determine the amounts of the ingredients that give a most acceptable product, i.e. with highest predicted score.

(c) From a similar analysis for the other response variable, which variables contribute significantly to crumbliness?

(d) Determine those conditions that produce cakes within the specification for crumbliness, i.e. with scores definitely between 4.0 and 7.0.

(e) What conditions would you recommend for Rudyard's New Rock Cakes?

11

Outliers and influential observations

11.1 Introduction

When a set of data has been obtained, it is always necessary to examine the observations for any that might be out of step with the bulk of the data. They could indicate a mistake in recording or calculation, a contaminated sample, or the effect of an unexpected change in conditions. Any of these causes would diminish the validity of the conclusions.

Two types of observation can distort an analysis: the rogue result or 'outlier', which has a value of a variable that is so different from the rest of the data as to be suspicious, and an 'influential' observation that, by being extreme in one or more of the independent variables, has an unduly strong pull on a regression line.

This chapter presents tests for outliers and influence.

11.2 An outlier in one variable

The most commonly used test for a single outlier in a set of data in one variable is Grubbs' test, which we illustrate with the following example:

Table 11.1 contains nitrosamine values obtained on samples from six batches of cured hams. They are also shown in the blob diagram of Figure 11.1.

Table 11.1 Nitrosamine values in 6 samples of ham

Sample	1	2	3	4	5	6	Mean	SD
Nitrosamine (mg/g)	57	48	91	53	56	61	61.0	15.32

Effective Experimentation: For Scientists and Technologists   Richard Boddy and Gordon Smith
© 2010 John Wiley & Sons, Ltd

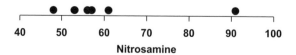

Figure 11.1 Blob chart of nitrosamine values.

We can notice that five of the results are in a narrow range and there is one very high result that may cause concern. The sample mean, 61.0, is equal to the highest of the remaining observations rather than being near the middle. Grubbs' Test is applied as follows:

Test value - $= \dfrac{|x-\bar{x}|}{s}$

where x is the suspected outlier, $\bar{x}$ is the sample mean (including the possible outlier),

and s is the sample standard deviation (including the possible outlier);

$= \dfrac{|91-61.0|}{15.32} = 1.96$

Table value -   From Table A.1 with 5 degrees of freedom: 1.89 at the 5% significance level.

Provided the assumptions underlying the test apply in this case, we can reasonably conclude (since the test value exceeds the table value) that a rogue result is present.

The assumptions of the Grubbs' test are that the underlying distribution of the data is normal. The validity of the test can only be judged from knowledge of the distribution from many previous results.

Let us take two hypothetical situations.

If the underlying distribution was normal, as in Figure 11.2, it is quite evident that a value of 91 is unlikely to have been obtained from this distribution. It would be correct to separate it from the rest of the data and investigate the cause. If, however, the underlying distribution was more like the skewed pattern in Figure 11.3, a set of data with five results between 48 and 61 and a sixth at 91 would be quite typical of this distribution and we would have no grounds for concluding that an outlier was present.

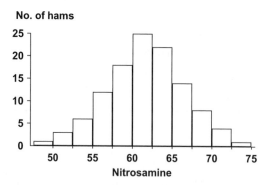

Figure 11.2 A histogram from an underlying normal distribution.

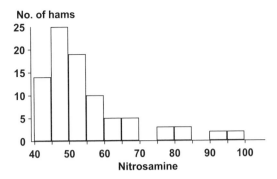

Figure 11.3   A skewed population.

So, what is an outlier?

An outlier is a result that does not fit the pattern (of errors) of the other observations.

We shall see it is important to talk about errors in situations with two or more variables. Since we are at the moment considering a single variable we can just look at the pattern of observations. It is clear from the definition and from looking at the two hypothetical distributions that in order to use an outlier test we must make a statement about the shape of the distribution. The most commonly used outlier tests assume the normal distribution, but others are available for skewed distributions. In all cases, the distribution must be stated in the hypotheses.

Previous data on nitrosamine levels were normally distributed, so it was appropriate to use Grubbs' test. It was discovered on investigation that there was a reason for the outlier: that side of ham had been left for the incorrect time in the curer. The value could therefore be excluded from the analysis (after reporting it separately). The remaining results had a mean of 55.0 and a standard deviation of 4.85, showing that an outlier can have some influence on the mean but a dramatic effect on the standard deviation.

11.3   Other outlier tests

There is another situation in which Grubbs's test will perform badly – the presence of two or more outliers. It is possible that in the presence of two outliers, Grubbs's test will fail to find even one! This is not too surprising since the alternative hypothesis refers to the most extreme and the test has been designed for situations in which only one outlier is present. There are other tests, for example:

   Table A.3 gives a test for two outliers – one high and one low – with the remainder of the observations coming from a normal distribution.

   Table A.4 gives a test for two outliers, with the remaining observations coming from a normal distribution, providing the outliers are either both high or both low values.

Thus we see that, to apply an outlier test, not only is it necessary to know the type of distribution but it is also necessary to know the number of outliers that can be expected and whether they occur in the same direction. The correct application of outlier tests is not an easy judgement.

11.4 Outliers in regression

In determining the validity of a regression equation that has been obtained, the assumption of a normal distribution of errors can be violated by the presence of an outlier.

An outlier in regression is indicated by an observation whose residual is unusually larger than the residuals of the other observations.

At Cleanwater Chemicals Ltd, a trainee statistician, Dan Lees, was given the task of investigating whether there was an effect of ambient temperature on the yield of a fermentation product. He obtained data from the plant manager on thirteen batches, and obtained a simple regression equation. The data, along with predicted values and residuals, are shown in Table 11.2, with a graph of the data and the regression line in Figure 11.4.

Table 11.2 Data and predicted yields (regression equation: $y = 48.0 + 0.795x$)

Batch	Temperature (x)	Yield (y)	Predicted yield	Residual
1	13	58	58.3	−0.3
2	18	59	62.3	−3.3
3	21	64	64.7	−0.7
4	17	65	61.5	3.5
5	13	62	58.3	3.7
6	22	68	65.5	2.5
7	29	69	71.0	−2.0
8	18	71	62.3	8.7
9	17	60	61.5	−1.5
10	16	55	60.7	−5.7
11	15	57	59.9	−2.9
12	12	55	57.5	−2.5
13	11	57	56.7	0.3
Mean	17.1	61.5		
SD	4.87	5.43		
% Fit = 51.0		Residual SD = 3.97		

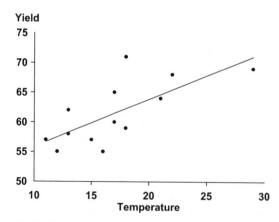

Figure 11.4 Regression line for yield on ambient temperature.

The residual standard deviation (RSD) was 3.97, but there is one observation with a residual of 8.7. It can clearly be seen well above the regression line.

Dan Lees performed an appropriate outlier test using the following test value:

$$\textbf{Test value} = \frac{\text{Maximum }|\text{Residual}|}{\text{RSD}\sqrt{1-\text{Leverage}}} \quad (\text{'Leverage' explained later})$$

$$= \frac{\text{Maximum }|\text{Residual}|}{\text{RSD}\sqrt{1-\dfrac{1}{n}-\dfrac{(X-\bar{x})^2}{(n-1)(\text{SD of }x)^2}}}$$

where X is the value of x for that observation, $\bar{x}$ is the mean of all the x values and n is the number of observations

$$= \frac{8.7}{3.97\sqrt{1-\dfrac{1}{13}-\dfrac{(18-17.1)^2}{12(4.87)^2}}} = 2.28$$

This test value is referred to as the 'studentized' residual.

Table value: From Table A.5, with a sample size (number of observations) of 13 and 2 coefficients in the equation (slope and intercept):

2.54 at the 5% level.

Although the point on the graph was sufficiently high above the line to attract attention, it is not far enough to be considered an outlier.

11.5 Influential observations

As he examined the graph in Figure 11.4, Dan Lees was concerned at the slope of the line. For all points except the extreme one on the right he felt that the line should be steeper, and he was disappointed that one observation alone could distort the relationship that appeared to apply to all the other observations.

Just as we can more easily open a door by pushing it at the furthest point from the hinges, so an observation has an unduly large amount of 'influence' or 'leverage' on a regression line if its value of the independent variable is at a considerable distance from the centre of the data.

With one independent variable, a point of influence can be detected from a graph, but it is necessary, as we shall see later, to have diagnostic statistics available, particularly with many independent variables.

The effect of each observation with independent variable equal to X is measured by the 'leverage':

$$\text{Leverage} = \frac{1}{n} + \frac{(X-\bar{x})^2}{(n-1)(\text{SD of }x)^2}$$

For the observation with $X = 29$,

$$\text{Leverage} = \frac{1}{13} + \frac{(29-17.1)}{12(4.87)^2} = 0.574$$

Before determining whether this observation has significant leverage, the following points should be made:

(1) Leverage refers only to values of independent variables. They measure the influence of the x values, not the responses.

(2) Leverage lies between 0 and 1.0. The higher the value, the more the leverage.

(3) The minimum value for leverage with only one independent variable is $1/n$, which occurs when an observation has the mean value of the independent variable.

(4) Average leverage is p/n, where p is the number of coefficients in the equation and n is the number of observations.

An approximate convenient 'table value' is $2p/n$.
With $p = 2$ and $n = 13$, the 'table value' would be 0.31.

As the observation being investigated had a leverage value of 0.574, Dan Lees can conclude that it has undue influence.

If that observation were omitted, the regression equation for the remaining observations would be

$$y = 44.9 + 0.994x$$

Dan's suspicions about the distorting effect on the slope were justified. Whatever linear relationship applied in a temperature range of $11\,^\circ$C to $22\,^\circ$C did not appear to continue to higher temperatures. The batch had been produced on a day of weather the like of which had not been seen for many years, so the value of the independent variable could be considered to be untypical.

11.6   Outliers and influence in multiple regression

The outlier test and measure of leverage are obtained by analogous methods of which the simple regression example earlier in the chapter was a special case.

Leverage in multivariate data, such as multiple regression, measures the influence in terms of the distance of the point from the centroid of the independent variables relative to the variability in the direction from the centroid to the point. For example, with two independent variables x_1, x_2 in Figure 11.5, although the points A and B appear to be the same distance from the centroid, B has much greater leverage (0.650) than A (0.352) because of the smaller variability in the data in the direction of B.

The formulae for studentized residual and leverage are necessarily much more complex for multivariate data, so are not given here.

In Chapter 9 we met an example in which ten experimental batches of an active ingredient of a chemical were made with the aim of developing a formulation with increased yield. The data in Chapter 9 had been 'cleaned up' after it had been discovered that one of the yields (batch 7) had been wrongly recorded. The original data is given here.

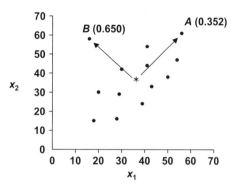

Figure 11.5   Influence in multivariate data.

The regression equation obtained was:

$$\text{Yield} = 108 - 9.07\,\text{weight} - 0.074\,\text{temperature}$$

The residuals and leverages are shown in Table 11.3.

Table 11.3   Residuals and leverages

Batch	Weight	Temperature	Yield	Residual	Studentized residual	Leverage
1	1.6	45	91.5	1.4	0.45	0.21
2	1.3	40	92.9	−0.3	−0.09	0.30
3	1.1	50	97.3	3.0	1.22	0.49
4	1.2	45	94.1	0.4	0.12	0.21
5	1.6	40	89.9	−0.6	−0.20	0.36
6	1.2	45	94.8	1.1	0.35	0.21
7	1.4	50	83.7	−7.8	−2.61	0.27
8	1.7	50	92.2	3.4	1.44	0.55
9	1.5	45	90.9	−0.1	−0.03	0.13
10	1.4	40	91.7	−0.6	−0.19	0.27

There are two observations that catch the attention: batch 7 has a very high studentized residual and the leverage for batch 8 is high.

From Table A.5, with 3 coefficients in the equation and a sample size of 10, the table value is 2.31. Batch 7, with a studentized residual of 2.61, is therefore an outlier. This batch was investigated. It was discovered that the yield had been wrongly recorded as 83.7, whereas it was actually 93.7. The correct value was used in the multiple regression analysis in Chapter 9.

The criterion for influence is $2p/n = 2 \times 3/10 = 0.60$.

Batch 8 with leverage of 0.55 does reach the 'table value' for an influential observation, but notice that its settings of weight and temperature are both at the extremes of the experimental design.

11.7   What to do after detection?

Once an outlier or influential observation has been detected, it should not automatically be discarded. It should first be checked to confirm that the data have been correctly calculated or recorded, and then investigated to ensure that there were no abnormal circumstances such as contamination or incorrect procedure.

Outliers should be reported, as they often reveal much information about the process, and the data reanalysed after discarding them.

Influential observations give a warning. They may tell something about a change in the relationship outside the domain of the rest of the data, but more often than not give a misleading suggestion of a relationship.

In the examples in this chapter the influential points distorted the relationship that applied to the rest of the data, so should be reported separately.

The statistics of the relationship should be obtained both with and without influential observations. If there is little change in the relationship, and the influential observation follows the same pattern as the rest of the data, it need not be discarded.

12

Central composite designs

12.1 Introduction

In Chapters 4 to 7 we have previously considered two-level factorial designs, which are extremely valuable in determining significant effects. In Chapter 10 we met three-level designs that can be used to fit response surface models and estimate optimum conditions for processes. We have seen that the confidence interval of prediction is at its lowest at the centre of the design but increases towards the extremes. If the optimum response occurred near the extremes of the design we would not have such a precise estimate of its location.

In this chapter we look at a design that is based on two-level and three-level designs, but that incorporates some additional features. The design uses replicated points at the centre of the design to improve estimation of curvature and to estimate true replicate error and thus enable the model error – that is the residual error after fitting the model – to be checked.

The central composite design is so planned that the precision of prediction is approximately the same throughout the experimental region. In addition, central composite designs are arranged in two blocks so that the experiment can be undertaken in two stages with a considerable time period between them.

12.2 An example: design the crunchiness

Jeremy Boyd-Smith, a research officer with Jackson Finefoods plc, is in the later stages of a development programme to produce a new crunchy biscuit. He has already managed to produce a biscuit with suitable properties but he realizes that improved crunchiness would lead to greater popularity. (Crunchiness is measured by the load required to snap the biscuit.) Boyd-Smith has established that two variables - the water content of the mix and the feed rate of the extruder – greatly affect the crunchiness. The feasible ranges for these variables have also been established and are given below:

Effective Experimentation: For Scientists and Technologists   Richard Boddy and Gordon Smith
© 2010 John Wiley & Sons, Ltd

Water content: 25% to 55%
Feed rate: 1.2 to 2.8 rev/s

He therefore wishes to investigate the effects of these variables in order to choose optimum conditions for both variables.

He assumes that the response will be affected by the levels of both water content and feed rate, with their quadratics and interaction, and if there is an optimum it could be anywhere within the experimental region.

He decides to use a central composite design, which can be conducted in two stages.

In the first stage Boyd-Smith carries out 7 trials using the same batch of raw materials. Perhaps wisely he decides not to investigate the extreme corners of the experimental region and he produces a design consisting of a 2^2 factorial with 3 extra points added at the centre. The design, together with the responses, is shown in Figure 12.1 and in Table 12.1. This is the **1st stage** of a central composite design.

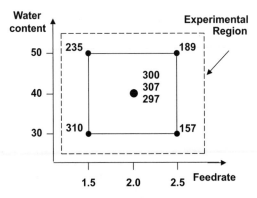

Figure 12.1   Design and results of the 1st stage.

Table 12.1   Results of the 1st stage

Trial	Water content (w)	Feed rate (f)	Crunchiness (y)
1	40	2.0	300
2	50	1.5	235
3	30	2.5	157
4	40	2.0	307
5	30	1.5	310
6	50	2.5	189
7	40	2.0	290

The design will enable Boyd-Smith to test the significance of the two main effects, the interaction of water content and feed rate, and curvature. Before doing so, he must obtain an estimate of variability.

12.3   Estimating the variability

There are two alternative methods of obtaining an estimate of the standard deviation:

(1) Using the replicate responses at the centre. These are:

$$300 \qquad 307 \qquad 290$$

These responses give:

$$\text{Mean} = 299$$
$$\text{Standard deviation} = 8.54, \text{ based on 2 degrees of freedom.}$$

We shall refer to this estimate of the standard deviation as the experimental SD.

(2) Using an estimate obtained externally from the experiment. Boyd-Smith has already obtained an estimate based on 16 batches made under identical conditions:

$$\text{Standard deviation} = 14.0 \text{ based on 15 degrees of freedom.}$$

We shall refer to this as the external SD.

Any estimate based on only 2 degrees of freedom is likely to be imprecise and so Boyd-Smith decides to use the external SD based on 15 degrees of freedom to assess the significance of the main effects. However, before proceeding, we should check whether the external SD is significantly different from the experimental SD using an F-test for the ratio of two standard deviations.

Test value: $= \dfrac{\text{Larger SD}^2}{\text{Smaller SD}^2}$

$= \dfrac{(14.0)^2}{(8.54)^2} = 2.69$

Table value: From Table A.6 (two-sided) with 15 degrees of freedom (for the larger SD) and 2 degrees of freedom (for the smaller SD): 39.43 at the 5% significance level.

Conclusion: There is no evidence to suggest that the variabilities are different.

12.4   Estimating the effects

We can now consider the main effects and interactions by compiling a two-way table of the responses at high and low levels of water content and feed rate (i.e. ignoring the replicated centre points). The marginal means and the means along the diagonals are also shown.

Feed rate	Water content 30	50	Mean	
				196.0
1.5	310	235	272.5	
2.5	157	189	173.0	
Mean	233.5	212.0		
				249.5

To test whether these effects are significant we shall use the two-sample t-test as outlined in Chapter 2.

$$\text{Test value} = \frac{|\bar{x}_A - \bar{x}_B|}{s\sqrt{\dfrac{1}{n_A} + \dfrac{1}{n_B}}}$$

where $\bar{x}_A, \bar{x}_B$ are the two means being compared, being the means of n_A, n_B observations respectively, and s is the residual standard deviation.

Main effect of water content (compare the column means):

$$\text{Test value} = \frac{|212.0 - 233.5|}{14.0\sqrt{\dfrac{1}{2} + \dfrac{1}{2}}} = 1.54$$

Main effect of feed rate (compare the row means):

$$\text{Test value} = \frac{|173.0 - 272.5|}{14.0\sqrt{\dfrac{1}{2} + \dfrac{1}{2}}} = 7.11$$

Interaction (compare the diagonal means):

$$\text{Test value} = \frac{|249.5 - 196.0|}{14.0\sqrt{\dfrac{1}{2} + \dfrac{1}{2}}} = 3.82$$

For each of those tests the table value from Table A.2 with 15 degrees of freedom is 2.13 at the 5% level. Thus, the interaction and the main effect of feed rate are significant, but not the main effect of water content.

Let us now turn our attention to curvature. Since all the central values are at the same point we can only establish whether any significant curvature is present, not which variables are contributing to it. To test for curvature we again use the two-sample t-test:

Mean of replicates at centre $= 299$ $(n = 3)$
Mean of corner points $= 222.75$ $(n = 4)$

$$\text{Test value} = \frac{|299.0 - 222.75|}{14.0\sqrt{\dfrac{1}{3} + \dfrac{1}{4}}} = 7.13$$

The table value is again 2.13. Clearly significant curvature is present.

12.5   Using multiple regression

Another approach for analysing this data is to use multiple regression. The intercorrelation matrix for the design is shown in Table 12.2.

Table 12.2   Intercorrelation matrix for the 1st stage

	w	f	wf	w^2	f^2
w	*	0.00	0.00	0.00	0.00
f		*	0.00	0.00	0.00
wf			*	0.00	0.00
w^2				*	1.00
f^2					*

The inter-correlations are ideal apart from the confounding of w^2 and f^2, which gives a correlation of 1.00. We will not be able to separate these terms so we must examine the two alternative equations and their associated response surfaces.

Since we know that the interaction and quadratic terms are significant we will proceed directly to the contour diagram for each equation, with the data or mean at each point added, shown in Figures 12.2 and 12.3.

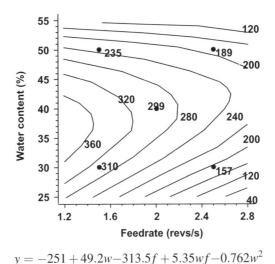

$$y = -251 + 49.2w - 313.5f + 5.35wf - 0.762w^2$$

Figure 12.2   Contour diagram with quadratic term in w.

There are several points to note about the contour diagrams. First, they are greatly different in shape. Figure 12.2 indicates that a maximum crunchiness will be obtained with 37% water and a feed rate of 1.2 rev/s, while Figure 12.3 indicates that 25% water should be used at a feed rate of 1.73 rev/s. Both of these conclusions cannot be right: one of them may be right, but the truth could well lie somewhere between them. We should also notice that at the experimental

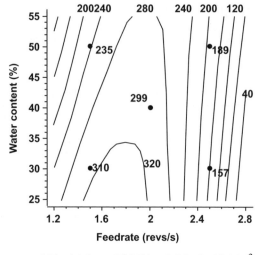

$$y = -251 - 11.7w + 906.5f + 5.35wf - 305.0f^2$$

Figure 12.3 Contour diagram with quadratic term in f.

points – the corners and the centre – the two diagrams give the same values, which are equal to the responses (or mean response) at these points. Finally, we must not forget that the contour diagram gives a spurious impression of great accuracy because it gives no indication of the confidence interval at each point.

Because it is important to obtain better estimates, Boyd-Smith decides to proceed to the 2nd stage of the design. In this stage 6 further trials are undertaken, 2 more at the centre and 4 at the 'star points' of the design. The full design is shown in Figure 12.4 and Table 12.3 with the results. The results for the first stage are in parentheses in Figure 12.4.

12.6   Second stage of the design

The intercorrelation matrix for this design is shown in Table 12.4. Notice we include another variable, a block variable B that takes a value of 1 for the first stage and a value of 2 for the second stage.

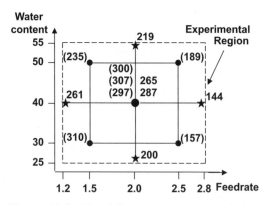

Figure 12.4 The full central composite design.

Table 12.3 Results from the full design

Trial	Water content (w)	Feed rate (f)	Crunchiness (y)	Block (B)
1	40	2.0	300	1
2	50	1.5	235	1
3	30	2.5	157	1
4	40	2.0	307	1
5	30	1.5	310	1
6	50	2.5	189	1
7	40	2.0	290	1
8	40	2.71	144	2
9	40	2.0	265	2
10	54.1	2.0	219	2
11	25.9	2.0	200	2
12	40	2.0	287	2
13	40	1.29	261	2

Table 12.4 Intercorrelation matrix for both stages

	w	f	wf	w^2	f^2	B
w	*	0.00	0.00	0.00	0.00	0.00
f		*	0.00	0.00	0.00	0.00
wf			*	0.00	0.00	0.00
w^2				*	-0.13	0.06
f^2						0.07
B						*

Table 12.4 illustrates an important feature of the design: the two stages have been chosen so that, should the stages give different levels of the response, no bias will occur in estimating the effects, i.e. the effects are not correlated with B.

We shall now proceed to obtain a response surface model by fitting all the terms, including the block variable, in a multiple regression analysis:

Constant: -241.10	% fit: 97.02	d.f.: 6	Residual SD: 13.99

	Variables in the equation			Variables available to add		
Variable	Coefficient	Decrease in % fit	Test value to delete	Variable	Increase in % fit	Test value to include
w	17.2968	0.08	0.41			
f	306.584	42.14	9.21**			
B	-19.0291	2.94	2.43$\sim$			
w^2	-0.3525	21.63	6.60**			
f^2	-152.874	25.97	7.23**			
wf	5.3500	7.27	3.82**			

We notice that the block variable is nearly significant. When blocks are not significant the term can be deleted from the equation.

Before looking at the contour diagrams we should examine the residual SD to see if it is reasonable relative to the replicate SD.

Replicate SD of stage $1 = 8.54$ based on 2 degrees of freedom

Replicate SD of stage $2 = 15.56$ based on 1 degree of freedom

$$\text{Combined replicate SD} = \sqrt{\frac{\sum(\text{d.}f. \times \text{SD}^2)}{\sum(\text{d.}f.)}}$$

$$= \sqrt{\frac{(2 \times 8.54^2) + (1 \times 15.56^2)}{2 + 1}}$$

$$= 11.37 \text{ based on 3 degrees of freedom}$$

We can now determine whether the residual SD is significantly greater than the replicate SD. [Should it be significant it will show that the model has a poor fit possibly due to an outlier or as a result of choosing the wrong form of model.]

To proceed, we first calculate sums of squares, which are explained in Chapter 17:

Sum of squares for model error plus replicate errors

$$= (\text{Residual SD})^2 \times \text{d.f.}$$
$$= (13.99)^2 \times 6$$
$$= 1174.3 \text{ based on 6 degrees of freedom}$$

Sum of squares for replicate error

$$= (\text{Replicate SD})^2 \times \text{d.f.}$$
$$= (11.37)^2 \times 3$$
$$= 387.8 \text{ based on 3 degrees of freedom}$$

By calculating the difference between the sums of squares we can obtain the model error sum of squares. The degrees of freedom are also obtained by subtraction.

Model error sum of squares

$$= 1174.3 - 387.8$$
$$= 786.5 \text{ based on } (6 - 3) = 3 \text{ degrees of freedom}$$

The ratio of the mean squares gives an appropriate test value:

Test value: $= \dfrac{\text{Model Error Mean Square}}{\text{Replicate Error Mean Square}}$

$$= \frac{(\text{Model Error Sum of Squares})/(\text{d.}f.)}{(\text{Replicate Error Sum of Squares})/(\text{d.}f.)}$$

$$= \frac{786.5/3}{387.8/3} = 2.03$$

Table value: From Table A.6, **one-sided** with 3 and 3 degrees of freedom:
9.28 at the 5% significance level.

The model error is not significantly greater than the replicate error, a conclusion that is clearly obvious from the similar magnitude of the two standard deviations. The model fits the data well.

Let us now examine the response surface. The full equation is

$$y = -241.1 + 17.30w + 306.6f + 5.35wf - 0.353w^2 - 152.9f^2 - 19.0B$$

By putting $B = 1$ we can now obtain a suitable equation for contour plotting. This is shown in Figure 12.5.

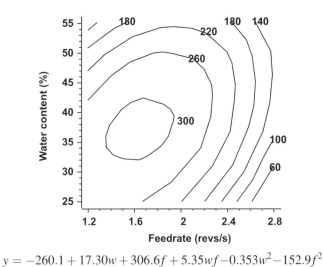

$$y = -260.1 + 17.30w + 306.6f + 5.35wf - 0.353w^2 - 152.9f^2$$

Figure 12.5   Contour diagram for equation estimated from both stages.

In Figure 12.5 the shape has changed considerably from the earlier diagrams and we now have a better estimate of the true response surface. It is interesting to note, however, that the values at the corner and centre are little changed from the previous designs. The star points have enabled us to obtain a better understanding of the rest of the surface.

What conclusions can Boyd-Smith draw from the analysis? The optimum appears to be at a feed rate of 1.6 and a water content of 37%. However, the confidence interval associated with the predicted crunchiness has a width of approximately ±26 and therefore there must be some uncertainty about the position of the optimum and the shape of the contour diagram. Boyd-Smith realizes that the determination of the optimum will require further investigation in a small region around the predicted optimum and because of the small changes in crunchiness in this region a large number of trials will be required for a successful conclusion to be reached. He believes this is impractical and therefore recommends that the plant should run at those conditions.

12.7 Has the experiment been successful?

Yes.

The experiment was designed so that knowledge could be gained about the relationship between crunchiness, water content and feed rate.

The investigation was designed as two experiments. This is good since it enables the strategy to be changed after the first experiment. However, it is important that the effects are not affected by any change in raw materials between experiments. This has been achieved.

The central composite design meant that the width of the confidence interval of prediction was similar throughout the experimental range. As the predicted optimum was in the middle of a quadrant, not at the centre of the space, the design had been helpful.

Knowledge was gained about the relationship through the space and a conclusion made about the position of the optimum.

But...

The confidence interval of predicted crunchiness was quite wide, and the crunchiness changed slowly in the region of the predicted optimum conditions. If the best conditions were required to be known more precisely, further experimentation, with a large number of trials, would be required in the region of the predicted optimum.

12.8 Choosing a central composite design

Table 12.5 defines a standard set of central composite designs for 2 to 6 variables. The table shows the number of points required at the corners, centre and star points together with the distance of the star points from the centre, assuming that the corners have been coded as ± 1. It should also be noted that fractional factorial designs are used to obtain the corner points with 5 and 6 variables. It is possible to use more than 2 blocks with the larger designs.

Table 12.5 Central composite designs

No. of variables	No. of observations				Total	Distance of star points from centre
	1st block		2nd block			
	Corner	Centre	Centre	Star		
2	4	3	2	4	13	1.414
3	8	3	3	6	20	1.682
4	16	4	3	8	31	2.000
5	16	3	3	10	32	2.000
6	32	4	5	12	53	2.378

12.9 Critique of central composite designs

These central composite designs have been produced using the following criteria:

(a) The confidence intervals at all points should have similar width.

(b) The effects should have zero correlations with each other.

(c) The blocking variable should have zero correlation with the effects, allowing the two stages to be carried out separately with a considerable time interval between them.

Central composite designs come very close to meeting these criteria. This does not, however, mean that the criteria are relevant to all experimentation. For instance, the first criterion results in the star points being placed further and further away from the centre as the number of variables increase. Is it really relevant to place the star points at 2.378 units from the centre in designs with 6 variables? Perhaps only two of the variables will be found to be significant – in which case 1.414 would give uniform confidence intervals. The main criterion in choosing the spread of points must be the experimental region of interest and **not** a desire for equal width of confidence intervals. We would recommend that the star points be limited to 1.414 (or 1.5 for a suitable rounded figure) no matter how many variables are included.

Another criticism of central composite designs is that they require a large amount of experimentation. With 6 variables, 53 observations are required, but if all the main effects, quadratics and interactions were significant – an extremely unlikely event – there would still be 17 degrees of freedom for the 'lack of fit' error. This may represent an undue waste of effort on the part of the experimenter and it may be better to use a quarter-fraction of the full factorial design, with centre points, for the first block.

One point of note: the term 'central composite design' is used to denote a type of design with the following characteristics:

- based on factorial designs;
- include centre and star points;
- usually arranged in blocks.

Thus, not all central composite designs satisfy all the specific design criteria defined above.

13

Designs for mixtures

13.1 Introduction

In many designs the range of variables such as temperature or pressure is limited only by physical constraints within the system. For a selected pressure, for example 1.7 atm, any temperature within its experimental range (for example 100–200 °C) can be chosen unless a combination at an extreme level is dangerous. When dealing with mixtures there is also a mathematical constraint for the mixture variables – the component proportions must sum to unity (or percentages must sum to 100%). Thus, for example, with a three-variable mix it is impossible to vary all three variables independently of each other and the mathematical constraint must be taken into consideration in both design and analysis. For example, if a batch of a fruit cocktail with orange, grapefruit and pineapple juice is mixed using 20% of orange and 40% grapefruit the amount of pineapple must be 40%. This restriction does not, however, apply to mixtures where the active ingredients are only a small proportion of a mix, the rest consisting of filler.

Shortly we will look at a practical case study involving concrete formulation but first let us examine the special features of the design and analysis of experiments involving mixtures.

13.2 Mixtures of two components

In Figure 13.1 we see that while two independent variables can lie anywhere within the square, proportions in a mixture must lie along the diagonal line which represents $x + z = 1$. The two-dimensional design space in Figure 13.1(a) has therefore been reduced to one dimension in Figure 13.1(b).

For a design to investigate a linear relation we need consider only two trials, one at either end of the experimental region, with x at 1 and z at 1.

Trial	Proportion of x	Proportion of z	Response (y)
1	1	0	10
2	0	1	20

Effective Experimentation: For Scientists and Technologists   Richard Boddy and Gordon Smith
© 2010 John Wiley & Sons, Ltd

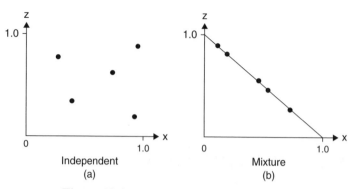

Figure 13.1   (a) Independent (b) Mixture.

Let us fit an equation of the form $y = b_1x + b_2z$.

(A regression equation associated with mixture variables does not include an intercept term as the intercept in a regression equation predicts the value of the response when all independent variables are equal to zero, which cannot exist in mixtures.)

Fitting the values of x, z and y into two simultaneous equations gives:

$$10 = b_1(1) + b_2(0)$$
$$20 = b_1(0) + b_2(1)$$

Therefore, $b_1 = 10; b_2 = 20$

and the regression equation is : $y = 10x + 20z$

The equation can be used to produce the response for any mixture, e.g.

For $x = 0.7, z = 0.3$ the response is

$$y = (10 \times 0.7) + (20 \times 0.3) = 13$$

If the model involved interactions and quadratics, the equation would be of the form:

$$y = b_1x + b_2z + b_3x^2 + b_4z^2 + b_5xz$$

For mixture variables with two components

$$x + z = 1$$

As a result we can substitute $x = 1 - z$ and $z = 1 - x$ into the above equation:

$$y = b_1x + b_2z + b_3x(1-z) + b_4z(1-x) + b_5xz$$

If we examine this equation there are only terms in x, z and xz. Thus, a quadratic term in an equation for a mixture is superfluous since it is represented by other terms. Thus, the equation for interactions and quadratics is usually of the form:

$$y = b_1x + b_2z + b_3xz$$

Because of the constraint on the levels of x and z any two terms in the original equation can disappear, but this is often the preferred form because of its symmetry and simplicity. Designs for a three-component mixture will be illustrated with a case study.

13.3   A concrete case study

Hardandfast produces ready-mixed cement to customers' requirements for hardness. They are considering a series of experiments to discover the effect of a number of variables on hardness with the aim of obtaining a more predictable product at lower cost.

Concrete is an extremely complex mixture in which such variables as the per cent of water and the proportions of sand, gravel and cement are important, as well as such process variables as particle size – usually represented by the source of supply – and batch size. However, we shall initially consider only the mixture components of sand, gravel and cement and return later to the effects of process variables. A three-variable mix can always be represented by a diagram (or lattice) on a triangular graph as shown in Figure 13.2.

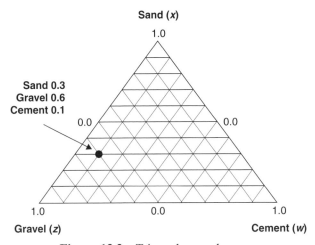

Figure 13.2   Triangular graph paper.

Each vertex represents one of the components at 1.0, or 100%. All along the opposite face that component is at 0.0.

If we plot any point within the triangle the percentages of all the components will always sum to 100. In Figure 13.2, for illustration, a mix with 10% cement, 30% sand and 60% gravel has been plotted. Thus, the triangular diagram is extremely useful for visualizing the required points in a design and, as we will see later, for contour diagrams of the predicted response surface.

However, experimentation across the full region of Figure 13.2 would be pointless. What is the good of concrete with 0% cement? Hence, the next step is to put some constraints onto the experimental region:

Sand must be at least 20%;

Gravel must be at least 50%;

Cement must be at least 10%.

We can now draw our constraints onto the triangular diagram as shown in Figure 13.3.

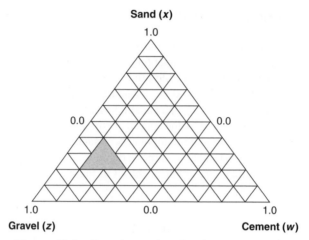

Figure 13.3   Experimental region for concrete mix.

This region is bounded by the following vertices:

$(S = 0.4, G = 0.5, C = 0.1);$ $(S = 0.2, G = 0.5, C = 0.3);$ $(S = 0.2, G = 0.7, C = 0.1)$

We can now expand the region onto a full diagram as seen in Figure 13.4:

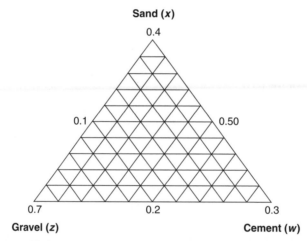

Figure 13.4   An expanded diagram of the experimental region.

To make the results more general we shall use coded values between 0.0 and 1.0, rather than the actual proportions when designing the experiment and analysing the results.

13.4   Design and analysis for a 3-component mixture

The design for a 3-component model which includes the 3 main effects and the 3 interactions is represented by the diagram in Figure 13.5.

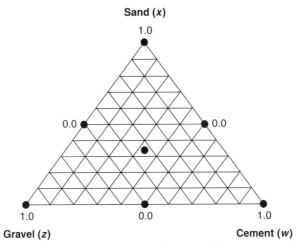

Figure 13.5   Experimental design.

It contains the three vertices, the midpoints of each face, and the centroid. (The three vertices are all that are needed for a model containing the main effects.)

For this example the design, in scaled form, is given in Table 13.1, together with the 28-day strengths in Newtons of the two batches that were made with each formulation.

Table 13.1   Design conditions and responses

Trials	Sand (x)	Gravel (z)	Cement (w)	Strength
1, 11	1	0	0	22.3, 23.5
2, 12	0	1	0	20.1, 23.8
3, 13	0	0	1	27.6, 26.6
4, 14	$1/2$	$1/2$	0	21.8, 24.8
5, 8	$1/2$	0	$1/2$	29.8, 27.0
6, 9	0	$1/2$	$1/2$	25.4, 24.1
7, 10	$1/3$	$1/3$	$1/3$	26.1, 23.9

The actual formulations are:

Trials	Sand	Gravel	Cement
1, 11	0.40	0.50	0.10
2, 12	0.20	0.70	0.10
3, 13	0.20	0.50	0.30
4, 14	0.30	0.60	0.10
5, 8	0.30	0.50	0.20
6, 9	0.20	0.60	0.20
7, 10	0.27	0.57	0.17

We can now proceed to fit a multiple regression equation to all 14 results using a forward stepwise procedure with **no constant** in the model but offering all two-component interactions for selection. The stepwise regression procedures are as follows:

Step 1:   Enter cement into the model

% fit	55.0
Test value	3.99
Table value (Table A.2, 13 d.f.)	2.16
Residual SD	17.3

Step 2:   Enter sand into the model

% fit	83.1
Test value	4.47
Table value (12 d.f.)	2.18
Residual SD	11.0

Step 3:   Enter gravel into the model

% fit	99.6
Test value	21.00
Table value (11 d.f.)	2.20
Residual SD	1.80

Step 4:   Enter the interaction of sand and cement into the model

% fit	99.8
Test value	2.58
Table value (10 d.f.)	2.23
Residual SD	1.46

Step 5:   No further terms are entered since the next term has a test value for the interaction of sand and gravel of 0.44, compared with a table value of 2.26.

Regression equation:

$$y = 27.1 \text{ cement} + 23.2 \text{ sand} + 22.2 \text{ gravel} + 12.2(\text{sand} \times \text{cement})$$

There are several points to note:

(a) The % fits are very high. This is usual when a constant is omitted from the equation.

(b) The residual SD is very high until all three components are included. This is not surprising in this example since all three components are well known to affect the strength of concrete.

(c) The trials with equal coded values of the three components could have been excluded from the design. They do, however, provide a valuable insight into the goodness of fit of the model to the data using predicted values and residuals. These are given in Table 13.2.

We see in Table 13.2 that the fit is just as good for the trials at the centroid as any other trial. The formulator is now in a position to use the equation to formulate a mix that gives the right proportions at minimum cost.

The regression equation can be represented by a contour diagram as shown in Figure 13.6.

We recall that axes are expressed in coded units, where the minimum and maximum of each component are scaled to 0.0 and 1.0.

The contour diagram is shown in relation to the actual values of the components in Figure 13.7.

Table 13.2 Predicted values and residuals

Sand	Gravel	Cement	Predicted strength	Residuals
1	0	0	23.2	−0.9, 0.3
0	1	0	22.2	−2.0, 1.6
0	0	1	27.1	0.5, −0.5
$\frac{1}{2}$	$\frac{1}{2}$	0	22.7	−0.9, 2.1
$\frac{1}{2}$	0	$\frac{1}{2}$	28.2	1.6, −1.2
0	$\frac{1}{2}$	$\frac{1}{2}$	24.6	0.8, −0.5
$\frac{1}{3}$	$\frac{1}{3}$	$\frac{1}{3}$	25.5	0.6, −1.6

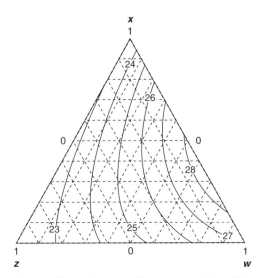

Figure 13.6 Contour diagram (coded units).

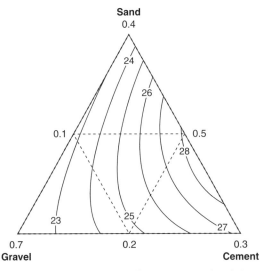

Figure 13.7 Contour diagram (actual units).

To interpret the contour plot, recall that the vertices correspond to the scaled value of a component equal to 1.0, i.e. the component being at its maximum value, and the opposite face having the scaled value of the component equal to zero, the component actually being at its minimum value.

The optimum conditions (those giving maximum strength for the concrete) are at $z = 0$, $w = 0.65$, $x = 0.35$, i.e. gravel 0.50, cement 0.23, sand 0.27.

13.5   Designs with mixture variables and process variables

Hardandfast now decide to extend the study to include two process variables – batch size and type of sand – which may affect the quality at time of use. They decide to include two levels of each variable.

For batch size, the low level is $10 \, \text{m}^3$ (coded '-1') and the high level $30 \, \text{m}^3$ ($+1$);

Two sand types may be used, as sand is obtained from one of two quarries, Fox Quarry (-1) or Badger Quarry ($+1$).

By coding these variables in that way their mean is zero. We can continue to have a regression model with no constant term. If we had coded them as, say, 1 and 2 we would have needed to include a constant term.

This gives a design with 3 mixture components and 2 process variables. They decide to carry out a full experiment by carrying out 4 trials (2 variables at 2 levels) at each of the seven mixture combinations used in the earlier experiment.

The design, together with the responses, are:

	Quarry		-1	$+1$	-1	$+1$
	Batch size		-1	-1	$+1$	$+1$
Sand	Gravel	Cement				
1	0	0	25.4	30.9	22.5	30.6
0	1	0	20.0	23.6	22.9	23.1
0	0	1	27.3	29.2	26.4	29.5
$1/2$	$1/2$	0	24.6	25.9	21.3	25.9
$1/2$	0	$1/2$	26.1	33.8	28.3	31.4
0	$1/2$	$1/2$	25.5	25.5	23.0	26.5
$1/3$	$1/3$	$1/3$	27.4	30.6	23.9	26.8

The experiment is analysed by offering all two-variable interactions (including those between process variables and mixture components) for inclusion in the stepwise regression equation. Again no constant is included. The stepwise regression procedure is as follows for those variables that are significant:

Step	Variable entered	% fit	Test value	d.f.	Table value	Residual SD
1	Cement	52.9	5.50	27	2.05	18.5
2	Sand	84.3	7.20	26	2.06	10.9
3	Gravel	99.1	20.43	25	2.06	2.64
4	Sand × quarry	99.6	5.99	24	2.06	1.71
5	Sand × cement	99.7	3.08	23	2.07	1.47
6	Quarry	99.8	2.62	22	2.07	1.31

The regression equation is -

$$y = 28.1\,\text{cement} + 26.8\,\text{sand} + 22.4\,\text{gravel} + 2.7(\text{sand} \times \text{quarry})$$
$$+ 10.3(\text{sand} \times \text{cement}) + 0.90\,\text{quarry}$$

Again we can represent the equation by using contour diagrams, but with the increased number of significant variables we need to use more than one diagram.

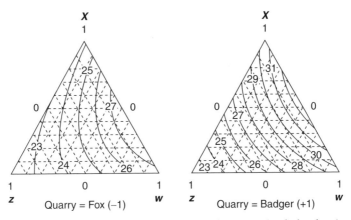

Figure 13.8 Contour diagrams for each quarry (coded values).

Figure 13.8 shows contour diagrams for two different coded levels of quarry.

The contours are shown again in relation to the actual levels of the components in Figure 13.9.

Notice the effect of the interaction−sand × quarry.

With sand at its lowest level (coded value equal to 0), the heights of the contours are little changed between diagrams. However, as the proportion of sand increases the height changes dramatically.

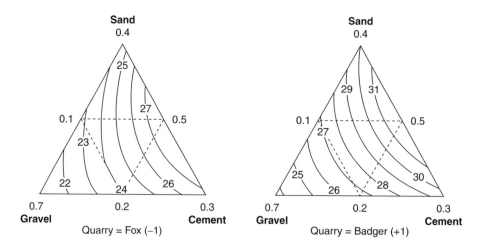

Figure 13.9   Contour diagrams for each quarry (actual values).

13.6   Fractional experiments

The number of trials needed in experiments involving both mixture and process variables increases rapidly as the number of variables increases.

No. of components	No. of trials	No. of process variables	No. of trials
2	3	2	4
3	7	3	8
4	15	4	16
5	31	5	32
6	63	6	64

Thus, a six-component mixture with 3 process variables would need $63 \times 8 = 504$ trials. Clearly a fractional design is required.

Let us look at an example of an experiment involving 3 components and 3 process variables. The eight trials needed for each mixture can be halved using $+ABC$ as the defining contrast for one mixture and then $-ABC$ for the next mixture to give the design in Table 13.3. (Refer to Chapter 6 for an explanation of the term 'defining contrast'.)

However, no such classical designs are available for obtaining a fraction of a mixture design. Therefore, it would be necessary to use a computer-assisted experimental design in which a number of candidate points is offered and the best selection is made. (Refer to Chapter 14 for the use of computer-aided experimental design.) Thus, for 6 components, up to 63 candidate points could be offered for selection from which the experimenter may require, say, only 16 trials. The computer will then search different combinations of 16 trials from the 63 candidates available and recommend the best selection. This also has the advantage that the experimenter can remove points that are not feasible. For example, a 6-component mix having 100% of one component may not be desirable.

Table 13.3 Design of $\frac{1}{2}$ fraction of a full experiment with 3 process variables

Mixture variables			Process variables	A	−1	+1	−1	+1	−1	+1	−1	+1
				B	−1	−1	+1	+1	−1	−1	+1	+1
				C	−1	−1	−1	−1	+1	+1	+1	+1
R	S	T										
1	0	0				X	X		X			X
0	1	0			X			X		X	X	
0	0	1				X	X		X			X
$\frac{1}{2}$	$\frac{1}{2}$	0			X			X		X	X	
$\frac{1}{2}$	0	$\frac{1}{2}$				X	X		X			X
0	$\frac{1}{2}$	$\frac{1}{2}$			X			X		X	X	
$\frac{1}{3}$	$\frac{1}{3}$	$\frac{1}{3}$				X	X		X			X

13.7 Was the experiment successful?

Yes.

Hardandfast designed a first experiment to examine the effects of the mixture variables – sand, gravel and cement – within a range of constraints on each component. They chose the best set of conditions to enable them to investigate a second model. They replicated each formulation and included a centroid point that enabled them to check the goodness of fit of the model. As a result they obtained a regression model for mixtures from which they could determine the optimum conditions, i.e. those that produced the hardest concrete and thus met the objective of the experiment.

The second experiment incorporated process variables. They crossed the mixture design with a factorial design. They concluded that one of the two quarries from which they could obtain the sand gave harder concrete.

But ...

In reality these may have been limited experiments. The mixture components were the crucial ones, so that their amounts were expressed as percentages of their total. Did the amount of added water make a difference? It was assumed that the same total weight of the components was the same and a fixed amount of water was added. However, it is well known that the amount of water is a critical component in concrete. It should have been included as a further variable.

14

Computer-aided experimental design (CAED)

14.1 Introduction

The classical experimental design – full factorial, fractional factorial, central composite . . . –
allows main effects and stipulated interactions and quadratics to be investigated with the effect
estimates not being biased by each other. This is achieved by obtaining zero intercorrelations
between all estimates. This is no mean achievement but it is obtained at a price – the design
must be as published and is, therefore, linked to specific situations and a designated number of
trials. There are many situations which will prevent the use of classical designs. Namely:

  (a) The amount of material or time means that we cannot complete a full design.

  (b) There are some combinations of conditions that are not acceptable. For example, in a
      chemical process high temperature and high pressure may be dangerous. In other
      processes, certain conditions will prevent the operation of the process.

  (c) A certain number of trials have already been completed and the experimenter now
      wishes to extend the trial.

  (d) Some favoured conditions need to be included in the trial. For example, a reference run
      is needed using present conditions and this run needs to be repeated several times
      during the trial.

We shall obtain the design using CAED, an Excel program developed by Statistics for
Industry. See the preface for access to the program.

14.2 How it works

Designs are about achieving information.

Effective Experimentation: For Scientists and Technologists   Richard Boddy and Gordon Smith
© 2010 John Wiley & Sons, Ltd

The two features of a good design are:

i. Low intercorrelations between effects including interactions and quadratics.

ii. Data points covering as wide a range as possible. This range not only applies to main effects but also to interactions and quadratics.

If we have a zero intercorrelation but a negligible range, the information will be minimal. If we have a large range but an intercorrelation nearing 1.0 the information about the effects will be minimal. Clearly we need to balance both criteria (i) and (ii) when designing an experiment. This is achieved by trying to optimize the determinant of the design matrix referred to as Det M. The higher the value of Det M the better. Classical designs usually give the maximum value of 1.0. Designs using CAED will achieve nothing like this value but we are not concerned about its actual value, only if it can be improved as the iteration proceeds. Let us look at an example.

14.3   An example

Perfect Paints Ltd have identified a problem with their paint that is making it less than perfect. The problem is that the paints are increasing their viscosity with storage time to such an extent that the paints are becoming unspreadable after three months. Titan Ox has been asked to investigate and has decided to examine four variables:

A	The amount of head space in the paint tin.
B	The solvent type (xylene or butanol).
C	The gas in the head space (air or nitrogen).
D	Speed of reaction of water being removed from the system.

Titan believes that there are likely to be interactions between the solvent type and the other three variables.

This gives the interactions:

$$AB, BC \text{ and } BD$$

Titan would also like to look at curvature for variables A and D.
Thus, he will investigate the following effects:

$$A \quad B \quad C \quad D \quad AB \quad BC \quad BD \quad A^2 \quad D^2$$

To investigate nine effects he will need a minimum of 10 trials, but in reality a few more trials are essential. He has been allotted sufficient time to carry out at least 11 trials but if the experiment runs well he could possibly obtain 16 trials.

Titan's next decision is to choose the level of the variables. Two of these – solvent type (xylene or butanol) and content of air space (air or nitrogen) are already fixed. The other two variables require at least 3 levels to allow the fitting of a quadratic. Titan chooses:

Amount of head space	2 ml, 4 ml and 6 ml
Speed of reaction	60, 70, 80 and 90 min

The speed of reaction is traditionally set at 10-min intervals and Titan therefore decides to use four levels to cover the desired 60 to 90 min. He also codes:

$$Xylene = 1 \qquad Butanol = 2$$
$$Air = 1 \qquad Nitrogen = 2$$

He is now in a position to start using CAED.

14.4   Selecting the repertoire

The user must offer the program a set of possible trials from which to select. This often requires much judgement but in this case we can offer the full set of combinations:

Head space value	Solvent	Head space content	Speed of reaction
2	1	1	60
4	2	1	60
6	1	2	60
2	2	2	60
4	1	1	60
6	2	1	60
2	1	2	60
4	2	2	60
6	1	1	60
2	2	1	60
4	1	2	60
6	2	2	60
2	1	1	70
4	2	1	70
.	.	.	.
.	.	.	.
.	.	.	.
6	2	2	90

There are a total of 48 cells ($3 \times 2 \times 2 \times 4$) in the repertoire.

14.5   Selecting the model and number of trials

The next stage is to select the model. This consists of all the coefficients that could be needed in a regression model. We thus ensure the 'constant' box is ticked and then tick:

$$A \qquad B \qquad C \qquad D$$
$$AB \qquad BC \qquad BD$$
$$AA \qquad DD$$

We can now enter the number of trials of 11 into the spreadsheet. This value must be above the number of 10.

14.6   How the program chooses the design set

The program will select randomly 11 out of 48 possible trials. It will then compute Det M and also the maximum correlation coefficient between effects (Max r). It will then delete 1 of 11 trials chosen randomly and substitute randomly one of the repertoire trials not in the design. Again, it will compute Det M and if this is a larger value than the previous Det M this will become the preferred design set. If not, it will be discarded. Each attempt to improve Det M is called an iteration. In total, 1000 iterations are carried out. This can be shown as a process map:

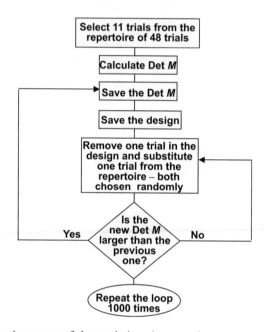

We can now examine some of the statistics given on the program:

Total Alternatives:	2E+10 or 20 billion is the total number of ways of selecting 11 trials from 48. Clearly not an option to pursue.
Repertoire:	The 48 trials we have designed.
Minimum Trials:	The number of effects, interactions and quadratics plus the constant.
Number of Trials:	11 – The number required in the design set.
Forced Trials:	0. We have not chosen to force any trials into our design set. If certain trials had already been carried out we would have forced these into the set.
Free Trials:	The number of trials – number of forced trials
Free Repertoire:	The number of trials in repertoire – number of trials in the design set
Number of Substitutes:	1. We have chosen to alter one trial at a time. In some circumstances it may be better to alter more than one at a time.

Det *M* and Max *r*:   The first 11 trials to be chosen give a Det *M* of 0.0006 and a Max *r* of 0.69. As the program proceeds one can see an alternative Det *M* being obtained and substituted into the best Det *M* when a higher value is obtained.

Better alternatives are produced, quickly at first and then very slowly - this is shown by the statistics ***Iterations since Change***. At the 1000th iteration the 'best' design set gives:

Det *M*	Max *r*
0.1302	0.16

We now have three alternative courses of action:

To accept the design set   With a Max *r* of 0.16 this would seem very reasonable.

To continue to iterate   Another 1000 iterations is unlikely to give a much better set.

To remove and then substitute   This may improve the set since it is possible that
2 trials in each iteration   certain combinations are needed in the design set.

Titan decides to substitute 2 at a time. However, no improved design can be found. Thus, we can stop the search. If an improved design had been found we should consider doing another 1000 iterations with 2 substitutes or increasing the substitutes to 3.

By now pressing the **Generate** button we can find our design matrix in the worksheet of the same name. The design is:

Trial	*A*	*B*	*C*	*D*
1	6	1	1	60
2	2	2	2	60
3	2	1	2	90
4	4	1	2	60
5	2	1	1	70
6	6	2	2	90
7	2	2	1	90
8	4	2	2	80
9	6	1	2	80
10	4	1	1	90
11	6	2	1	60

The design is best illustrated graphically using two-way plots:

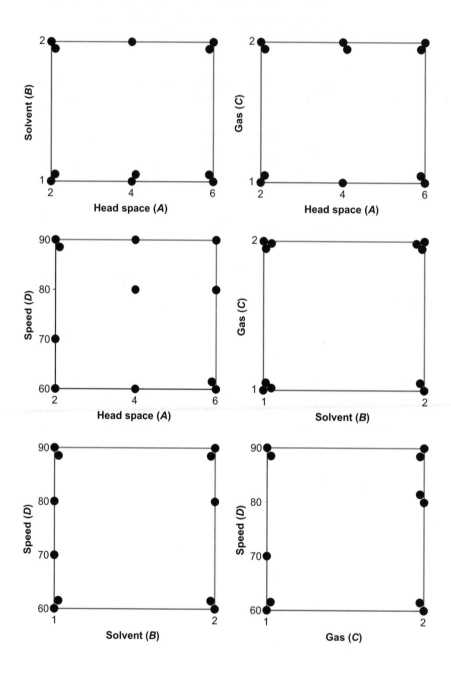

The design can also be evaluated by an intercorrelation matrix, which is given below:

	B	C	D	AA	AB	BC	BD	DD
A	0.00	0.00	−0.16	0.00	0.00	0.00	0.16	0.00
B		0.10	0.04	0.15	0.00	0.10	0.04	0.15
C			0.10	−0.15	0.00	−0.10	−0.04	−0.15
D				−0.06	0.16	−0.04	−0.01	−0.06
AA					0.00	−0.15	−0.06	0.08
AB						0.00	−0.16	0.00
BC							0.10	−0.15
BD								−0.06

By any standards this intercorrelation matrix is superb.

The number of trials now can be increased one by one to obtain a design ranging from 11 to 16 trials. In order to do this it is necessary to **force** the chosen set into the design and then increase the number of trials by one:

Number of trials	Det M	Max r
11	0.1302	0.16
12	0.1415	0.25
13	0.1397	0.28
14	0.1335	0.30
15	0.1301	0.32
16	0.1440	0.25

The extra trials have the following design:

Trial	A	B	C	D
12	4	2	1	70
13	2	1	2	60
14	6	1	2	90
15	6	2	1	90
16	6	2	2	60

It is interesting to note that Det M does not always increase. The change in value in Det M can be evaluated as follows:

Increasing the number of observations will always give better information. However, if the Det M substantially decreases there will be little gain in information. On the other hand, if the Det M substantially increases there will be considerable gain of information. Titan would be well advised to carry out 12 rather than 11 trials but should not go to greater lengths to do 16 rather than 15 trials.

The other statistic, **Max r**, is not as important as Det M since we can see that many alternatives give lower values but higher Det M. This is because they are not fully exploiting the range of the variables.

14.7 Summary

So how do we choose a design? The first important element is to choose the effects, interactions and quadratics to be investigated. The design set will totally reflect the choice – put an interaction into the model and it will treat it just as importantly as a main effect when searching for a design set. Thus, it is necessary to be realistic with possible interactions. The same applies to quadratics but even more so since the inclusion of quadratics will necessitate a far bigger repertoire.

Choosing the repertoire is also of paramount importance. There are several rules:

i) If an effect variable is present and if only a linear effect is required (i.e. no quadratics) only two levels are required since the design will only choose the extreme values.

ii) If a quadratic is required it is optimum to have only 3 levels with the central level mid-way between the two extremes.

iii) The simplest repertoire to choose is all possible combinations but this usually leads to a large repertoire that may exceed the specifications of the program. Large repertoires also cause inefficiency in using the program. If possible, use your knowledge of fractional factorials to limit the size of repertoire.

iv) If some data points will always be chosen, e.g., the central point in 3-level designs – these should always be forced into the design set. This will improve the efficiency of the iterations.

v) With some designs, e.g., central composites, it is efficient for some certain points to be repeated. If this is likely, a point should be entered more than once into the repertoire and then forced into the design set.

Having stipulated the model and repertoire we should follow the procedure as outlined in the chapter but always be willing to try different alternative procedures to ensure that the 'best' or near best design has been chosen. The time spent designing the experiment is minimal compared with executing it!

14.8 Problems

1. An R&D chemist is carrying out an investigation into the tensile strength of rubber. He wishes to design an experiment to assess the curvature due to carbon black content (A) and % natural rubber (B) together with the **linear** effect of catalyst concentration (C). He also wishes to assess **all** the interactions. He has enough material for at least 11 trials and possibly 12, 13 or 14 trials. A summary of the variables and their ranges is given below.

Carbon black (A):	20 to 40
% of natural rubber (B):	50 to 90
Catalyst concentration (C):	2.0 to 3.0

(a) Design a suitable experiment using CAED for 11, 12, 13 and 14 trials. Advise the chemist on a suitable number of trials.

(b) He now decides to assess curvature of all three variables. Design suitable trials for 11, 12, 13 and 14 trials. Design suitable experiments.

2. A development chemist wishes to investigate the effect on yield of temperatures between 200 °C and 300 °C and pressures between 10 and 16 atm. Unfortunately, certain combinations of pressure and temperature are considered unsafe. The safe region is shown below:

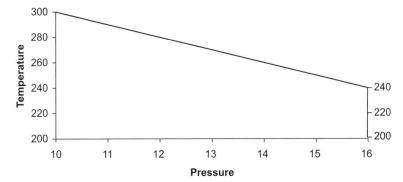

(a) Use CAED to obtain a design to investigate linear, quadratic and interaction effects in 14 trials – a careful choice of repertoire is needed.
(b) Repeat the analysis extending the repertoire by introducing duplicate points at the four 'corners' and their central point.

15

Optimization designs

15.1   Introduction

This book is mainly concerned with the design and analysis of experiments that lead to knowledge of how a response varies across an experimental region, or the comparison of the means of data from two or more conditions.

In this chapter, however, we are concerned solely with determining the set of conditions that lead to the optimum response, for example the highest yield or lowest impurity, without necessarily gaining knowledge of the relationships.

Three methods that are often used to optimize processes are:

(a) Using **response surface methods**. Carrying out a designed experiment and using the responses to fit an equation involving linear, quadratic and interaction effects. Obtaining a response surface and then estimating the optimum.

(b) Carrying out an **EVOP ('evolutionary operation') trial** in which a number of small changes are made to process variables based on a factorial design. The trials are repeated until significant effects are obtained. The process is moved in the direction of the optimum and the procedure repeated.

(c) Using a **simplex** approach in which the number of trials at each step is kept to a minimum and new process conditions are chosen irrespective of whether the observed changes are significant or not.

In this chapter we will concentrate on EVOP and simplex methods. We shall not cover response surface methods since they were covered in Chapter 10 but will compare and contrast all three methods at the end of the chapter.

15.2   The principles behind EVOP

We shall start by looking a case study.

Effective Experimentation: For Scientists and Technologists   Richard Boddy and Gordon Smith
© 2010 John Wiley & Sons, Ltd

Diazamine is the trade name for an organic chemical produced by Westman Chemical Corporation. The product was the culmination of a long and expensive research programme followed by development trials on a pilot plant. It is now produced on a short-run batch process and a number of runs have successfully taken place – successful as regards the operation of the process and meeting quality specification of the product, less successful on meeting financial criteria such as the net return on the cost of the research and development programme. The poor return is mainly due to the failure to obtain the expected yield. Westman have stated that they are unwilling to put extra research into the process and expect the plant operating personnel to improve the yield. To this end, the plant manager decides to run an Evolutionary Operation (EVOP) programme. The principles behind EVOP are as follows:

1. Variables that it is believed affect yield should be identified.

2. These variables should be perturbed slightly so that their effect can be estimated but the plant should continue to produce within specification.

3. When a variable(s) has been shown to be significant the plant should operate at the improved conditions.

The first step is to decide the most likely variable to affect yield. After consulting the Research Department, the plant manager decides that the two most likely variables to affect yield are weight and 'time to addition'. Figure 15.1 shows the place of these variables in the production process.

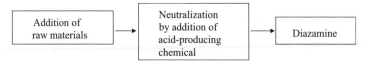

Figure 15.1   Flow chart of production process.

The reaction is completed by the addition of the acid-producing chemical and it is the time to this addition and the weight added that are the chosen variables for investigation.

15.3   EVOP: The experimental design

The plant manager's next step is to examine production records for the last 30 batches. From these he obtains:

> Operating conditions: Weight added 120 kg, Time to addition 75 min
>
> Mean yield = 562 kg per batch;   Standard deviation = 10.1 kg

After consultation he decides to perturb the weight added by ±5 kg and the time of addition by ±5 min.

The next step is to choose an experimental design.

The most efficient design for a two-variable investigation is a 2^2 factorial experiment. However, added to this design is a batch that is produced with conditions mid-way between the two levels for each variable, that is with conditions at the centre of the design. Since the suggested levels are 70 and 80 min for time and 115 and 125 kg for weight, the conditions for the reference batch are 75 min and 120 kg. The full design is shown in Figure 15.2.

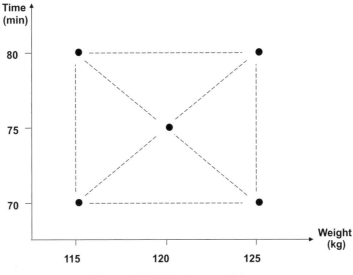

Figure 15.2 Design conditions.

The plant manager runs a batch at each of the conditions and obtains the yields shown in Table 15.1.

Table 15.1 Yields for first cycle

Weight	Time	Yield
115	80	567
115	70	581
120	75	573
125	70	580
125	80	573

Using the methods introduced in Chapter 4 to analyse a two-level factorial experiment, the **effect** of weight is the difference between the mean yield at 125 kg and the mean yield at 115 kg:

$$\text{Effect of weight} = \frac{580 + 573}{2} - \frac{567 + 581}{2}$$
$$= 2.5 \text{ kg}$$

A 90% confidence interval is given by Effect $\pm \dfrac{2ts}{\sqrt{n}}$

where n is the number of trials used in the calculation, which is 4 (the reference batch at the centre plays no part in this calculation);

s is the batch-to-batch standard deviation based on the last 30 batches $= 10.1$;

t from Table A.2 with the same degrees of freedom as s (29 degrees of freedom) at a 90% confidence level $= 1.70$. (The reasons for choosing a 90% confidence level will be explained later.)

$$90\% \text{ confidence interval} = 2.5 \pm \frac{2 \times 1.70 \times 10.1}{\sqrt{4}}$$

$$= 2.5 \pm 17.2$$

Similarly, the effect of time and the interaction can be computed.

$$\text{Effect of time} = \frac{580 + 573}{2} - \frac{567 + 581}{2}$$

$$= -10.5 \text{ kg}$$

To compute the interaction let us show the factorial results in a 2×2 table:

Time	Weight		
	115	125	573.5
70	581	580	
80	567	573	
			577

Interaction $=$ mean yield on the leading diagonal$-$mean yield on the secondary diagonal

$$= 577 - 573.5 = 3.5 \text{ kg}$$

Lastly, we can check for curvature by obtaining a confidence interval involving the centre conditions.

90% confidence interval for curvature:

$$(\text{Centre point} - \text{Mean of outer points}) \pm ts\sqrt{\frac{5}{n}}$$

$$= (573 - 575.2) \pm \left(1.70 \times 10.1 \sqrt{\frac{5}{4}} \right)$$

$$= -2.2 \pm 19.2$$

Thus, the summary of results from the EVOP design is:

Weight	2.5 ± 17.2
Time	-10.5 ± 17.2
Weight $\times$ time	3.5 ± 17.2
Curvature	-2.2 ± 19.2

Thus, none of the effects are significant, which is not too surprising since the conditions were only perturbed slightly. However, the principle behind EVOP is that the conditions are repeated — referred to as a cycle – until a significant effect is obtained, or the programme is aborted. In this instance, 3 cycles were needed as shown in Figure 15.3.

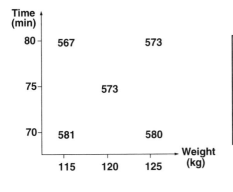

Effect	Estimate
Weight	2.5 ± 17.2
Time	-10.5 ± 17.2
Weight $\times$ time interaction	3.5 ± 17.2
Curvature	-2.2 ± 19.2

Cycle 1

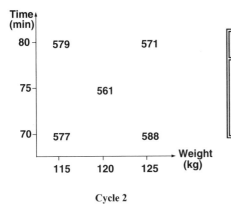

Effect for sum of 2 cycles	Estimate
Weight	2.0 ± 12.1
Time	-9.0 ± 12.1
Weight $\times$ time interaction	-3.0 ± 12.1
Curvature	-10.0 ± 13.6

Cycle 2

Effect for sum of 3 cycles	Estimate
Weight	-0.5 ± 9.9
Time	-10.8 ± 9.9
Weight $\times$ time interaction	-3.5 ± 9.9
Curvature	-4.8 ± 11.1

Cycle 3

Figure 15.3   Cycles 1 to 3.

We observe that as more cycles are run the confidence interval decreases because there are more observations. Thus, after 3 cycles they are in a position that the time of addition is a significant variable and this phase of the EVOP programme is therefore complete. They therefore will move the standard conditions for time of addition from 75 min to 70 min but will leave the weight at 120 kg. If the plant manager decides to institute a new phase he has several choices:

(a) Perturb time of addition by ±5 min around the new conditions.

(b) Since weight was not significant, the perturbation may have been too small. Therefore, perturb weight by ±10 kg.

(c) Exclude weight and introduce another variable.

Let us now look at two features of the analysis:

The use of 90% confidence intervals gives a 1 in 10 chance of making a wrong decision. Although the chance of a wrong decision is somewhat higher than usual it is necessary because of the small perturbations to the conditions. Also, if a wrong decision does occur, it is likely to be reversed at the next phase.

The curvature term is necessary to check whether the standard conditions are at the optimum. Should the curvature be the only term to be significant it would indicate that an optimum has been reached and the programme should be discontinued.

15.4   Running EVOP programmes

We have focused on a case involving one response. In practice, all other parameters will need to be in specification so it is essential that the product from all conditions remains within specification. This is a major consideration when choosing the amount of perturbation to each variable.

The EVOP programme will need substantial training of operating personnel. Instead of a quality philosophy of 'when in control – leave alone' operators will be asked to change conditions at each batch. Thus, they must understand the importance of those changes and be informed with a suitable visual presentation of how the programme is proceeding.

15.5   The principles of simplex optimization

Simplex optimization also seeks to determine the optimum conditions by a sequence of moves within the experimental space.

A selection of conditions is initially made so the experimental design is in the shape of a simplex (a triangle in two dimensions, a tetrahedron in three dimensions). One of the chosen initial conditions would be the present operating conditions.

After the batches have been run, the conditions of the least successful trial are replaced by the its reflection across the opposite side or face.

This procedure continues until the results are similar and the design converges on one set of conditions.

One of the criticisms of EVOP is the amount of time required before moving to better operating conditions. Simplex overcomes this difficulty at the expense of not checking for significance and therefore often moving in the wrong direction. However, such a move will be quickly reversed and is therefore not a fundamental objection to the method.

15.6   Simplex optimization: an experiment

Let us look at how Simplex can be applied using the Diazamine process as an example involving only two dimensions. Such an example is shown in Figure 15.4. In practice more dimensions will be used.

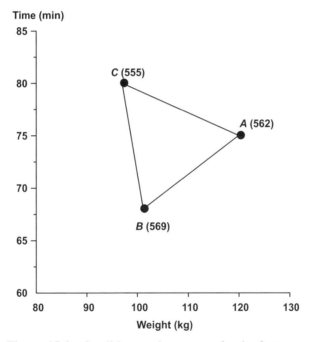

Figure 15.4   Conditions and responses for the first step.

One of the points relates to the present operating conditions: 120 kg and 75 min. The other two sets of conditions are equivalent to points at the vertices of an equilateral triangle. This effect can be achieved by suitable scaling of the axes of the graph. The responses for the three points (i.e. the yields at each set of conditions) are also given on the graph. The design and responses are given in Table 15.2.

Table 15.2 Conditions and responses for the first step

Trial	Weight	Time	Yield
A	120	75	562
B	101	68	569
C	97	80	555

The next step in the sequence is to take the point with the lowest response and reflect it about the opposite side of the triangle. This is shown in Figure 15.5.

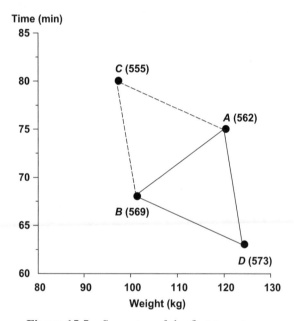

Figure 15.5   Summary of the first two steps.

We see the new point, D (124 kg, 63 min), is chosen so that it is perpendicular to AB and is the same perpendicular distance from AB as is the point C. We can also see that in order to complete the new triangle we need only one further trial. This is denoted by D and on running the plant a yield of 573 kg is obtained.

The next step uses triangle ABD. We choose the vertex with the lowest response, namely A, and reflect it about the axis BD. Thus, the procedure will gradually reach an optimum although the error in the observations might occasionally result in a move in the wrong direction. However, a number of difficulties need to be considered:

(a) Rogue values. An unduly high value will be retained in the simplex and the points will be reflected around it. To avoid this possibility it is good practice to repeat conditions pertaining to a value that has remained in the simplex for $k + 1$ trials, k being the number of variables. Thus, for two variables any value that remains in three successive simplexes is removed and the trial repeated.

(b) The simplex may be near a saddle-point for the variables under consideration but well away from its optimum if further variables are included. To avoid this, further variables may need to be included. This is easily achieved by setting the new variable at a different level for the next point under consideration.

(c) The chosen range of the process variables may be too small, resulting in the procedure returning to the same simplex after several intermediate steps. (Theoretically, the procedure would find the optimum but it would take far too many steps for it to be worthy of practical consideration.) In this case it is good practice to double the length of the reflection, to the point D', as shown in Figure 15.6.

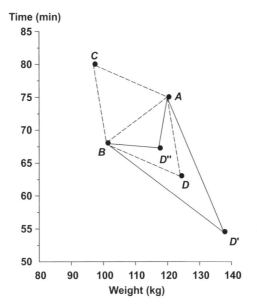

Figure 15.6   Changing the length of the reflection.

Thus, the perpendicular distance from D' to AB is twice that of C to AB. All future reflections will use the size of the new triangle. Note it is not necessary to rescale the variables to make the triangles equilateral; we can reflect about the midpoint of each side.

(d) The range of the process variables in the simplex may be too large, resulting in the simplex being reflected backwards and forwards across the same axis as the method keeps 'overshooting the mark'. In this case it is usual to halve the length of a reflection, to the point D", in a similar manner to that used to double the length as shown in Figure 15.6.

15.7   Comparison of EVOP, simplex and response surface methods

Each of the three methods has strengths and weaknesses and it is important to be aware of these when choosing a method. Let us look at these in detail.

15.7.1   Number of trials (batches) required

EVOP makes only small perturbations and therefore requires a large number of trials. In fact, one of the problems with EVOP is that management become impatient for results while EVOP is not showing any success. With Simplex, there is a higher chance element in the method so it can require few or many trials. On the other hand, response surface methods are highly efficient, using fractional designs, and therefore require fewer trials.

15.7.2   Production of in-specification product

If EVOP is run correctly all product will be within specification. This is one of the strengths of EVOP. No such guarantee can be given to Simplex since its procedure takes account of only one response and ignores other parameters. Response surface methodology relies on a 'bold' choice of conditions in order to obtain information quickly, so it would be unlikely that the resulting product would meet specification.

15.7.3   Obtaining optimum

EVOP will obtain an optimum slowly while Simplex will obtain it far quicker providing the present operating conditions are well away from the optimum. Response surface methods are, however, only efficient in reaching the region of optimum and the estimate of an optimum will be poorer than by the other methods.

15.7.4   Understanding variables

Because EVOP consists of a series of designed experiments it should give a good understanding of important variables and interactions. Simplex, on the other hand, will give a badly designed experiment when subjected to multiple regression analysis, and will give no information about the importance of variables. On the other hand, response surface methods will give excellent information about the nature of the variables.

15.7.5   Amount of judgement required

EVOP requires a fairly high level of judgement regarding important variables and the level of their perturbations that will keep the product in specification yet lead to an optimum. Simplex is a far more automatic process once the variables and their initial ranges are chosen. Response surface methods need a high level of judgement because of the fractional nature of the designs.

15.7.6   Coping with nuisance variables

In an experimental programme there are many variables, extraneous to the object of the experiment, which can change. For example, these can include feed material, operators and humidity. Such variables will tend to increase the variability within an EVOP programme, thus making the method less efficient.

Simplex copes badly with nuisance variables since changes in these variables can result in wrong reflections. Response surface methods should accommodate such variables without difficulty provided the researcher has shown judgement about them when designing the experiment.

16

Improving a bad experiment

N.B. The reader should read Chapters 9 and 14 to gain the full benefit of this chapter.

16.1 Introduction

In Chapter 9 we saw how a research scientist, Dr Ovi Dhose, struggled to analyse the results of an experiment because of poor experimental design. He was trying to determine the effects on the yield of a manufacturing process of five variables:

B: pH of the primary reaction;

C: Weight of acid added to the secondary reaction;

D: Age of catalyst in the secondary reaction;

E: Temperature of the secondary reaction;

F: Reaction time of the secondary reaction.

He ran ten batches, which are shown with the yields in Table 16.1.

He found problems caused by high intercorrelations between some of the independent variables. Dr Dhose now realizes that both temperature and time could have a curved relationship – that is why they were included at three levels. Their quadratics, denoted by E^2 for temperature and F^2 for time, need to be added to the analysis.

The full intercorrelation matrix is shown in Table 16.2.

Det *M* for this design is 0.00001. For an explanation of Det *M* see Chapter 14.

There are two correlations that would cause concern in an experimental design: 0.97 between *B* and *C* and –0.92 between *D* and F^2. The reasons can be seen in Figures 16.1 and 16.2.

The points on the graph nearly lie on a straight line, indicating the high correlation between *B* and *C*. If we are to reduce the intercorrelation we must have points in both shaded regions.

Effective Experimentation: For Scientists and Technologists   Richard Boddy and Gordon Smith
© 2010 John Wiley & Sons, Ltd

Table 16.1 Results from the experiment

		Reaction				
		Primary	Secondary	Secondary	Secondary	Secondary
Batch	Yield	pH	Weight of acid (kg)	Catalyst age	Temperature (°C)	Reaction time (min)
	A	B	C	D	E	F
1	91.5	9.8	1.6	1	45	70
2	92.9	9.2	1.3	2	40	70
3	97.3	8.8	1.1	3	50	70
4	94.1	9.0	1.2	4	45	50
5	89.9	9.8	1.6	5	40	50
6	94.8	9.0	1.2	6	45	50
7	93.7	9.6	1.4	7	50	50
8	92.2	10.0	1.7	8	50	60
9	90.9	9.6	1.5	9	45	60
10	92.9	9.2	1.4	10	40	60

Table 16.2 Intercorrelation matrix

	B	C	D	E	F	E^2	F^2
B pH	*	0.97	0.20	0.07	−0.06	0.10	−0.35
C Weight		*	0.26	−0.07	0.00	0.11	−0.44
D Age			*	0.04	−0.44	0.14	−0.92
E Temp				*	0.00	0.00	0.00
F Time					*	0.15	0.27
E^2						*	−0.03
F^2							*

The points on the graph show a strong curved relationship. This must be broken to obtain a low intercorrelation by introducing points in the shaded regions.

Dr Dhose now wishes to run further trials but he does not have a free hand in choosing conditions since pH is determined by the primary reaction and the catalyst age must be 11 since it will be the 11th batch produced with this batch of catalyst. He could extend his design without using a program but it would be difficult because other conditions need to be chosen to keep the low correlations between other variables.

He decides to use a computer-aided experimental design program (CAED), as described in Chapter 14, and stipulates that the ten conditions already used are included in the design. The next batch has a pH of 9.9, so this is included in the repertoire shown in Table 16.3.

To determine the best 11th batch, the program forces in the original ten batches and considers all the options on the repertoire. It chooses the highlighted one as the best.

This leads to the intercorrelation matrix shown in Table 16.4.

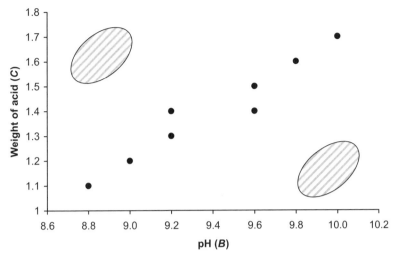

Figure 16.1   Levels of pH and weight of added acid.

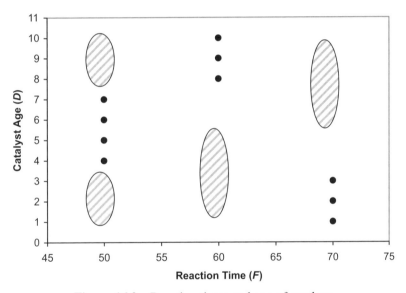

Figure 16.2   Reaction times and age of catalyst.

This is clearly an improvement. The correlation between B and C has been reduced from 0.97 to 0.66 and between D and F^2 from 0.92 to 0.84. Det M has also improved to 0.0005.

Dr Dhose can now proceed to add a further batch to his experiment. The next batch has a pH of 9.2 and the age of the catalyst must be 12. These values are changed in the repertoire, the

Table 16.3 Additional points in repertoire in order to choose 11th batch

pH	Weight	Age	Temperature	Time	
B	C	D	E	F	
9.9	1.1	11	40	50	
9.9	1.1	11	45	50	
9.9	1.1	11	50	50	
9.9	**1.1**	**11**	**40**	**60**	< - Chosen conditions
9.9	1.1	11	45	60	
9.9	1.1	11	50	60	
9.9	1.1	11	40	70	
9.9	1.1	11	45	70	
9.9	1.1	11	50	70	
9.9	1.7	11	40	50	
9.9	1.7	11	45	50	
9.9	1.7	11	50	50	
9.9	1.7	11	40	50	
9.9	1.7	11	45	60	
9.9	1.7	11	50	60	
9.9	1.7	11	40	70	
9.9	1.7	11	45	70	
9.9	1.7	11	50	70	

Table 16.4 Intercorrelation matrix for design with 11 batches

	B	C	D	E	F	E^2	F^2
B pH	*	0.66	0.34	−0.07	−0.04	0.18	−0.44
C Weight		*	−0.01	0.10	−0.02	−0.01	0.20
D Age			*	−0.15	−0.36	0.24	−0.84
E Temp				*	−0.01	−0.09	0.15
F Time					*	0.15	−0.09
E^2						*	−0.18
F^2							*

values of the remaining variables staying the same as before. The chosen trial is at

pH = 9.2, weight = 1.7, age = 12, temperature = 45, time = 70. Det M = 0.0058

This continues for the next two batches to give:

pH = 9.7, weight = 1.1, age = 13, temperature = 45, time = 70. Det M = 0.0108
pH = 9.0, weight = 1.7, age = 14, temperature = 40, time = 50. Det M = 0.0153

This set of 14 trials give the intercorrelation matrix in Table 16.5.

Table 16.5   Intercorrelation matrix for design with 14 batches

	B	C	D	E	F	E^2	F^2
B pH	*	0.25	0.10	0.03	0.07	0.07	−0.43
C Weight		*	0.16	−0.04	−0.11	0.06	−0.07
D Age			*	−0.21	−0.10	0.00	−0.31
E Temp				*	0.11	−0.17	0.09
F Time					*	−0.17	0.00
E^2						*	−0.23
F^2							*

We see that all the correlations are below 0.50 and in particular the correlation between pH and weight is only 0.25 instead of the original 0.97.

We see also the improvements in Det M and the largest correlations as shown in Table 16.6. The increases in Det M means that we are obtaining considerably more information from each additional trial.

Table 16.6   Values of Det M and maximum correlation

Number of trials	Det M	Maximum correlation
10	0.0001	0.97
11	0.0005	0.84
12	0.0058	0.61
13	0.0108	0.45
14	0.0153	0.43

Multiple regression can now proceed without any problems and should lead to a far more definite conclusion than was reached in Chapter 9.

16.2   Was the experiment successful?

Yes.

A bad experiment in Chapter 9 had been rescued with four extra batches, conditions for which have been chosen to optimize criteria for good experimental design.

But...

Clearly there are technical issues. Should the catalyst age have been extended from 10 to 14 or should a new catalyst have been started at batch 11? Also, the recommended weight of added acid was extreme, leading to an under- or overdose. This could lead to batches being abandoned.

17

How to compare several treatments

17.1   Introduction

In Chapter 2 we met an experiment that was planned in order to test whether a change in a process led to an improvement in the product. That is a very common situation in experimental science, but often we may have several formulations, suppliers, or 'treatments' that we wish to compare in order to decide:

(i) are there significant differences among them?

(ii) which one is the best?

17.2   An example: which is the best treatment?

Gordon Golightly is in charge of quality at a major filament lamp manufacturer. For some time he has been concerned about the poor performance of certain batches of lamps and has traced this to the high oxide levels on the surface of the tungsten filament. Golightly has now turned his attention to the cleaning process of the filaments in which coils of wire are slowly transported through a cleaning medium, chemical A. He wishes to investigate the performance of four other possible cleaning media – chemicals B and C and wet and dry hydrogen – to see whether there are differences, are they better than chemical A and which is the best?

Golightly therefore plans an experiment. The design needs to ensure than all media occur in a fair way throughout the experiment and there is no bias on any of them, either by the others or by any extraneous influences. To carry out the investigation he selects 20 coils and randomly assigns four to cleaning by each medium. The order of testing of the 20 coils is also random, as Golightly realizes that it would not be a sensible order to test all A, then B, etc. in case there was any underlying trend or step change in the testing equipment.

This experimental design is known as completely randomized.

Effective Experimentation: For Scientists and Technologists   Richard Boddy and Gordon Smith
© 2010 John Wiley & Sons, Ltd

The results of his investigation are shown in Table 17.1.

Table 17.1 Oxide levels after cleaning by each medium

Treatment (medium)	Chemical A	Chemical B	Chemical C	Wet hydrogen	Dry hydrogen
Oxide (ppm)	90	67	63	87	95
	92	81	70	78	98
	84	67	90	85	78
	94	75	69	80	99
Mean	90.0	72.5	73.0	82.5	92.5
SD	4.3	6.8	11.7	4.2	9.8

Overall mean $= 82.1$

The nature of the data is such that it is appropriate to analyse it using analysis of variance.

17.3 Analysis of variance

There is variability in data, arising from measurement error, sample-to-sample variability, batch-to-batch variability, and deliberately introduced differences caused by using different treatments or processes in an investigation.

Analysis of variance (ANOVA) will separate the variability from different causes.

To illustrate ANOVA in the present example the data are shown in a graph in Figure 17.1.

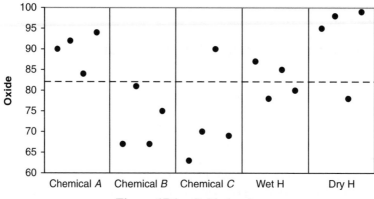

Figure 17.1 Oxide levels.

The total variation in the data can be seen from how much the points vary above and below the overall mean, indicated by the dashed line at 82.1.

The variation in the data is due to two main causes:

variation between coils in the same treatment arising from process variability and measurement error;

differences that may be expected because different treatments have been included in the experiment.

ANOVA will determine how much of the total variation in the data is attributable to each of these causes. The resulting table is shown in Table 17.2.

Table 17.2 Analysis of variance table

Source of variation	Sum of squares	Degrees of freedom	Mean square	Test value
Due to treatments	1382.8	4	345.7	5.45
Residual	951.0	15	63.4	
Total	2333.8	19		

We are interested in the differences between treatments.

The two sources of variation are 'due to treatments', which we would expect since the experiment was designed to compare them, and the remainder, referred to as 'residual'. The treatments were chosen to be put into the experiment to create differences, but there is no deliberate reason that coils in the same treatment would be different. Theirs is just the remaining unexplained variation.

The sum of squares column represents the sum of squared deviations of the data from a mean.

The total sum of squares is the sum of the squared deviations of all the data from the overall mean. Each deviation can be calculated from Table 17.1 or seen in the vertical distances of the points from the dashed line in Figure 17.1.

Total sum of squares = 2333.8.

The 'residual' sum of squares is the sum of squared deviations of each observation from the mean of its treatment. Each deviation can again be calculated from Table 17.1 or seen in the vertical distances of the points from each treatment mean in Figure 17.2.

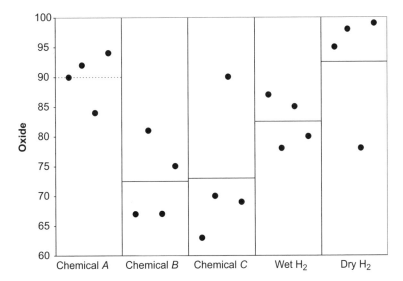

Figure 17.2 Oxide levels and treatment means.

Residual sum of squares $= 951.0$.

As we know that the total sums of squares is divided into the two causes, the 'due to treatments' sum of squares can be obtained by difference. It does, however, represent the variation of the treatment means about the overall mean.

'Due to treatments' Sum of Squares $= 2333.8 - 951.0 = 1382.8$.

It is always worthwhile checking the degrees of freedom.

There were 20 observations, so total degrees of freedom are $20 - 1 = 19$.

We are comparing 5 treatments, so degrees of freedom for treatments is $5 - 1 = 4$

Residual degrees of freedom are therefore $19 - 4 = 15$, but also because there are $4 - 1 = 3$ degrees of freedom from the four observations in each of the five treatments, so $3 \times 5 = 15$.

The next step in the analysis is to determine whether the treatments are significantly different using a significance test known as the F-test or variance ratio test.

Null hypothesis -   The treatment population means are equal.

Alternative hypothesis -   The treatment population means are not all equal.

(Notice that we cannot imply that every mean is different from every other one. Under the alternative hypothesis there will be some differences.)

The test value will compare the mean square between treatments and the residual mean square.

If there are no differences between the true means for the treatments, the mean square for treatments is not expected to be zero, but is expected to be the same as the residual. (Even if there are no differences caused by the treatments the observed means cannot be equal because of the residual error.)

Test value -

$$\frac{\text{Mean square between treatments}}{\text{Residual mean square}}$$

$$= \frac{345.7}{63.4}$$

$$= 5.45$$

(This is known as the F-ratio, or Fisher's variance ratio).

Table values -   From Table A.6 (one-sided) with 4 and 15 degrees of freedom:
                      3.06 at the 5% significance level;
                      4.89 at the 1% significance level.

Decision -   We reject the null hypothesis.

Conclusion -   The treatment means are different. Since the test value is greater than the table value at the 1% level the means are significantly different at a 1% level of significance.

At this point we have completed analysis of variance, and have concluded that there are differences between the means of the treatments.

If the variation between treatments had not been significant, we would have stopped at this stage.

We have determined that the means are significantly different. This is a necessary first step in the analysis, but for Golightly it is only part of the analysis, since his main aim is to determine **which** treatments give significantly lower oxide levels. This will be achieved using a multiple comparison test.

17.4   Multiple comparison test

The first step is to arrange the treatment means in order of magnitude as shown in Table 17.3:

Table 17.3 Treatment means arranged in ascending order

Treatment	Mean
Chemical B	72.5
Chemical C	73.0
Wet hydrogen	82.5
Chemical A	90.0
Dry hydrogen	92.5

The next step is to obtain an estimate of variability.

$$\text{Residual SD } (s) = \sqrt{\text{Residual mean square}}$$
$$= \sqrt{63.4}$$
$$= 7.96 \text{ based on 15 degrees of freedom}$$

The third step is to determine the least significant difference (LSD), the smallest difference between two means that would be significant at the 5% level.

This is obtained from -

$$5\% \text{ LSD} = ts\sqrt{\frac{1}{n_A} + \frac{1}{n_B}}$$

where s is the residual SD, t from Table A.2 at a 5% significance level has the same degrees of freedom as s and n_A, n_B are the numbers of observations in the two treatments being compared.
$s = 7.96$, t (5% significance level, 15 degrees of freedom) $= 2.13$, $n_A = n_B = 4$

$$5\% \text{ LSD} = 2.13 \times 7.96\sqrt{\frac{1}{4} + \frac{1}{4}}$$
$$= 12.0$$

Thus, any pair of treatments whose means differ by more than 12.0 are significantly different.

Thus, the difference in means between chemical B and chemical C is only 0.5 and therefore the two chemicals are not significantly different, whereas the difference between chemical B and dry hydrogen is 20.0 and these two treatments are significantly different.

With 5 treatments there are 10 different pairs. A simple way of showing all these comparisons is shown in Table 17.4, where those means that are **not** significantly different from each other are bracketed together.

Table 17.4 Significant differences between treatment means

Treatment	Mean
Chemical B	72.5 ⌉
Chemical C	73.0 ⎤
Wet hydrogen	82.5 ⌋⌉
Chemical A	90.0 ⎤
Dry hydrogen	92.5 ⌋

Thus, for example, dry and wet hydrogen are not significantly different but dry hydrogen is significantly different from chemical C.

Clearly Golightly has now completed the analysis of his experiment and as in most experiments the conclusions are not completely clear cut, but considerable knowledge has been gained. He can now see that chemicals B and C will both give a significantly better performance than chemical A, and that neither wet nor dry hydrogen give any improvement. It is unclear whether chemical B or C (or indeed wet hydrogen) is the best.

The above test is one of many multiple comparison tests. It is the simplest but has one major drawback: there is a 5% chance that each pair of treatments will be declared significantly different when they are the same. This is acceptable with a single pair, but with many comparisons could lead to a number of pairs being wrongly declared significant. To overcome this problem a number of other multiple comparison tests have been devised, the most popular one being Tukey's HSD (honestly significant difference).

17.5 Are the standard deviations significantly different?

One further aspect of the investigation is to check whether the standard deviations are consistent. There are many different tests for standard deviations all using different hypotheses. We shall demonstrate these using Cochran's test.

17.6 Cochran's test for standard deviations

Null hypothesis -   The population SDs are equal.
Alternative hypothesis -   The highest SD comes from a different population from the rest of the SDs.

Test value -

$$= \frac{(\text{Largest SD})^2}{\sum (\text{SD})^2}$$

$$= \frac{(11.7)^2}{(4.3)^2 + (6.8)^2 + (11.7)^2 + (4.2)^2 + (9.8)^2}$$

$$= 0.435$$

Table value -	from Table A.7 with 5 groups (treatments) and 4 observations (coils) in each group: 0.60 at the 5% significance level.
Decision -	We cannot reject the null hypothesis.
Conclusion -	There is not enough evidence to suggest that the highest SD is different from the rest of the SDs.

Thus, Golightly concludes the variability in impurity amongst coils is not due to different treatments but is due either to the coils themselves or to the test method.

If Cochran's test gave a significant conclusion, what should we do?

If one SD stands out from the rest it may be because one treatment contains an outlier, which should be tested.

17.7   When should the above method *not* be used?

When the coils have not been distributed randomly between the treatments.

When the coil-to-coil variability or test error is so great that it will mask any treatment effect. For Cochran's test, when the errors are skewed.

17.8   Was Golightly's experiment successful?

Yes.

He has compared the five treatments.
He was able to eliminate two of the candidate cleaners.

He has been able to conclude that there are differences and that two treatments give a significant improvement on the current one. He can now either do a further experiment to differentiate between the three treatments or choose one of them based on other factors such as cost.

But...

The conclusions were based only on one response variable, which was clearly important. Perhaps the effects on other response variables need to be looked at.

17.9   Problems

1. In order to reduce the corrosion to the inner surface of tins of polish, a manufacturer is experimenting with different surface coatings applied to the tin. Altogether, 6 different coatings are under investigation and three tins have been tested with each type of coating. An assessment of the corrosion is made visually on a scale of 1 (low) to 10 (high) at the end of the trial. Altogether five assessors are used and the values given below are the means of the five assessors:

Surface coating (Treatment)	A	B	C	D	E	F
Corrosion	5.8	4.8	4.0	3.2	5.2	6.4
	7.8	6.8	4.4	4.2	6.4	5.0
	6.8	7.0	4.8	4.0	4.6	6.0
Mean	6.8	6.2	4.4	3.8	5.4	5.8
Standard deviation	1.00	1.22	0.40	0.53	0.92	0.72

(a) Fill in the sources of variation and degrees of freedom in the table below:

Source of variation	Sum of squares	Degrees of freedom	Mean square
	To be calculated		To be calculated
	To be calculated		To be calculated
	To be calculated		

Using **Effective** enter the data into the ANOVA sheet.
Check your degrees of freedom in (a) against the table in the spreadsheet.
(b) Carry out a significance test to determine whether the treatment means are significantly different.
(c) Carry out a multiple comparison test and indicate the significant differences between treatments.

2. The laboratories of Getwell Private Hospitals plc conduct trials each year on a range of antibiotics before awarding a year's contract to the supplier of the most effective antibiotic.

Three solutions of each drug, of identical concentration, were prepared, and inoculated with *Staphylococcus aureus* in the concentration of 10^6 organisms/ml. Viable counts were determined after storage for 8 hours, and the log densities were as shown in the table below.

Log bacterial density of *S. aureus* with 4 antibiotics

Drug	A	B	C	D
	3.52	4.73	5.68	4.31
	4.18	4.29	4.83	4.92
	3.25	3.91	4.47	4.71
Mean	3.65	4.31	4.99	4.65
SD	0.48	0.41	0.62	0.31

(a) Fill in the sources of variation and degrees of freedom in the table below:

Source of variation	Sum of squares	Degrees of freedom	Mean square
	To be calculated		To be calculated
	To be calculated		To be calculated
	To be calculated		

Using *Effective* enter the data into the ANOVA sheet.

Check your degrees of freedom in (a) against the table in the spreadsheet.

(b) Carry out a significance test to determine whether the means for the antibiotics are significantly different.

(c) Carry out a multiple comparison test and indicate the significant differences between treatments.

(d) Which is the most effective antibiotic?

18

Experiments in blocks

18.1 Introduction

In Chapter 17 we met an experiment that was designed in order to establish whether there were significant differences between several treatments. The inclusion of different treatments was the only controlled or explainable source of variation. The samples within each treatment were assumed to be identical.

It is not always possible for experiments comparing the effects of different treatments to be carried out in this simple manner. The following situations are typical examples:

(a) Comparison of the yields of different processes may have to be split between several batches of raw material.

(b) Sensory testing of formulations usually has to be undertaken using several panellists who may have different sensory thresholds for the characteristic assessed.

(c) Trials of the effect on growth of animal diets may be influenced by differences in size and metabolism of the animals.

In each of these situations there is another source of variation in addition to variation between treatments. The experimenter will not be interested in the magnitude of variation in the yields contributed by using different batches of material, or the effect of panellist-to-panellist or animal-to-animal variation, but it will be there, and he will need to design his experiment so that he can quantify it and take account of it in the analysis otherwise the experiment could be ineffective.

18.2 An example: kill the sweat

Freshair Personal Products are developing a new antiperspirant, Nosweat, and have now selected three formulations – NSX, NSY and NSZ – whose effectiveness they wish to test and compare.

Effective Experimentation: For Scientists and Technologists   Richard Boddy and Gordon Smith
© 2010 John Wiley & Sons, Ltd

All the chemical and instrumental testing has been completed. They now wish to test the products on real people.

The measurement they will use is 'percentage reduction in sweating'.

They first thought of treating one group with NSX, another with NSY and a third with NSZ, but realized that there would be person-to-person variation in levels of perspiration and the effective of an antiperspirant. They have therefore designed an experiment in which each of a number of 'subjects' would be treated with all three formulations.

Five subjects were recruited. The effect of each formulation of antiperspirant was measured on each subject, the order being randomized. The results are displayed in Table 18.1.

Table 18.1   Percentage reduction in sweating for three formulations of antiperspirant

| | \multicolumn{5}{c}{Subject} | |
Formulation	A	B	C	D	E	Mean
NSX	20	29	32	25	27	26.6
NSY	29	34	38	29	39	33.8
NSZ	23	33	35	24	30	29.0
Mean	24.0	32.0	35.0	26.0	32.0	29.8

Notice the layout of the data, which reveals certain aspects of the design of the experiment:

(a) each formulation was tested five times, once on each of the subjects;

(b) the same five subjects were used for all formulations.

An experiment planned in such a way is called a 'randomized block' experiment. The name originated from its use in horticultural trials where the yield of plants would depend on the condition of the soil, and the field used for a trial would be divided into areas or blocks within each of which the soil changed little, although there might be differences (clay, water, sand) over the whole field. The differences between blocks were not of interest in themselves, but constituted variation that needed to be accounted for. Within each block all experimental treatments were used in random order of location.

If we inspect Table 18.1, we see that much of the variation between the three results for a formulation can be attributed to systematic differences between the subjects. Some subjects consistently respond more to an antiperspirant than other subjects. The subjects are the 'blocks' in this experiment.

18.3   Analysis of the data

If the experiment had not been planned to take account of the subject-to-subject differences (the 'blocks'), the results might have been analysed by one-way analysis of variance as in Chapter 17. This is shown in Table 18.2. It would be concluded that the differences between formulations were not significant.

Table value (Table A.6, one-sided, 2 and 12 degrees of freedom) $= 3.89$ at the 5% significance level.

This would be a surprising conclusion. If we examine Table 18.1 again we can see a remarkable degree of consistency between the results from the different subjects. NSX is the

Table 18.2 One-way analysis of variance

Source of variation	Sum of squares	Degrees of freedom	Mean square	Test value
Between formulations	134.4	2	67.2	2.82
Residual	286.0	12	23.8	
Total	420.4	14		

least effective formulation for all subjects except D, while NSY is the most effective for all subjects. Analysis by one-way analysis of variance was unable to detect these consistent differences between formulations because the 'residual' sum of squares included the contribution from differences between subjects, and it therefore swamped any differences between the formulations. To estimate the true residual variability, we need to calculate the 'between subjects' or 'blocks' variation and isolate it in the analysis.

Degrees of freedom for 'between subjects' = number of subjects − 1 = 4.

In both the 'sum of squares' and 'degrees of freedom' columns, the contributions due to 'between-subjects' are removed from 'residual'. The correct analysis of variance table is given in Table 18.3.

Table 18.3 Correct analysis of variance table

Source of variation	Sum of squares	Degrees of freedom	Mean square	Test value
Between formulations	134.4	2	67.20	17.01
Between subjects (blocks)	254.4	4	63.60	
Residual	31.6	8	3.95	
Total	420.4	14		

Notice that in Table 18.3 the amount of variation attributable to differences between formulations is unchanged from Table 18.2. The residual mean square, representing the amount of variability that cannot be explained, has been reduced from 23.8 to 3.95 by removing the variation that is present between subjects. This is the true measure of residual variability in such an experiment.

The test value has increased from 2.82 to 17.01, which is now significant at the 1% level, although the number of degrees of freedom has reduced.

(Table value from Table A.6 (one-sided) with 2 and 8 degrees of freedom
= 4.46 at the 5% level; 8.65 at the 1% level).

A multiple comparison test then uses the new residual mean square:

$$\text{Residual SD}(s) = \sqrt{3.95} = 1.99 \text{ with 8 degrees of freedom}$$

$$5\% \text{ least significant difference} = ts\sqrt{\frac{1}{n_A} + \frac{1}{n_B}}$$

where

s is the residual SD, t is a coefficient from Table A.2 at the 5% significance level with the same degrees of freedom as s

and n_A, n_B are the numbers of observations in the formulations being compared.

$$t = 2.31, \ s = 1.99, \ n_A = n_B = 5$$

$$5\% \, \text{LSD} = 2.31 \times 1.99 \sqrt{\frac{1}{5} + \frac{1}{5}} = 2.90$$

The mean effect of each formulation is shown in Table 18.4.
Formulations bracketed together are not significantly different.

Table 18.4 Formulation means

Formulation	Mean
NSX	26.6⎤
NSZ	29.0⎦
NSY	33.8

It is concluded therefore that NSY is significantly more effective than either NSX or NSZ, while there is no significant difference between NSX and NSZ.

Notice that the randomized block design is a special case of a two-way experiment (see Chapter 19), where instead of having two variables of interest there is one variable of interest and the 'block' variable.

18.4   Benefits of a randomized block experiment

The experiment is designed in such a way that the observations fall into homogeneous units or 'blocks', within which all treatments occur.

The variation between blocks can be quantified and extracted from what would have been assumed to be 'residual' in a one-way analysis.

Comparisons between treatments are thus more efficient than if blocking had not been used.

18.5   Was the experiment successful?

Yes.

The experiment successfully took account of the subject-to-subject variation and made more sensitive comparisons of the formulations than would have been possible if the experiment had not been designed in 'blocks'.

But ...

The experiment was inadequate. For such an experiment many more than five subjects should have been recruited. With a larger panel of subjects it is likely that the difference between NSX and NSZ would also have been significant.

18.6   Double and treble blocking

Randomized block and full factorial experiments are designed so that the effects of two or more possible contributors to differences in the data can be examined simultaneously. An experiment may involve trials at every combination of chosen conditions of, say, amount of an ingredient and baking temperature; or single or replicate measurements may be made on several tasters who have been presented with samples of different formulations of a foodstuff.

There is sometimes the need to include further factors in an experiment. By doing so, we might be making the required number of trials impracticable and we would need to consider a fractional design. It is often used when the additional factor cannot be incorporated in a full factorial design (for example it is an 'order' effect) and an alternative approach is required. The Latin Square design is one solution to such a situation involving three factors, provided each factor has the same number of levels.

Following the thinking in the randomized block experiment, the recipes are the variable of interest; 'dogs' and 'days' are both block variables.

This design can be used for three variables of interest, but it is a very parsimonious design – three variables each with three degrees of freedom and an assumption that there are no interactions.

18.7   Example: a dog's life

Best Friend Foods are a small independent manufacturer of foods for pets with their own successful niche in the market despite the efforts of two major competitors. Johan Thor, a food scientist in their small and highly innovative R & D department, plans an experiment to compare the acceptability to dogs of two of Best Friend's recipes (*A* and *B*) and one from each of their main competitors, *C* and *D*.

Four collies are selected from the research kennels. The experiment is to be designed so that each dog will be fed a different recipe on each of four days. In case there is any effect on the acceptability of a recipe of the previous day's recipe (a 'carry-over' effect), each dog will receive the recipes in a different order that will be recorded.

There are therefore three sources of variation, or factors, in the experiment:

4 recipes;

4 dogs;

4 days.

It is not possible to arrange a full factorial experiment, as, for example, each dog can only have one recipe 'first'; it cannot subsequently have another recipe 'first'.

18.8   The Latin square design

A Latin Square design is used, so that each dog has a different order of the four recipes and each recipe occurs once in each place in chronological order. The design is a form of fractional factorial, a $^1/_4(4^3)$.

Johan Thor's first attempt to devise such a plan is in Table 18.5, showing the recipe to be fed to each dog on each day.

At first Johan is pleased with this design. The design is nicely balanced; each dog has each recipe once and each recipe occurs once in each day. It satisfies the requirements for a Latin

Table 18.5 First attempt at Latin Square design

Dog	Day			
	Tuesday	Wednesday	Thursday	Friday
1	A	B	C	D
2	D	A	B	C
3	C	D	A	B
4	B	C	D	A

Square design. There is one problem in the orders of the recipes. If we look at recipes A and B, for example, we see that B occurs the day after A for three dogs, the only exception being when B occurs on the first day. The same problem occurs with B and C, C and D, and D and A. If there were a carry-over effect of one recipe it will affect the same recipe every time. It would be better if the antecedents were more varied and the carry-over effect were balanced across the recipes.

Fortunately, Johan was able to find an improved design, the Williams Latin Square, when he consulted the literature. It is shown in Table 18.6.

Table 18.6 A better Latin Square design

Dog	Day			
	Tuesday	Wednesday	Thursday	Friday
1	A	B	C	D
2	B	D	A	C
3	C	A	D	B
4	D	C	B	A

With this design the balance is maintained, but in addition each recipe is preceded (on the day before) by a different one of the other recipes or none at all. Johan decides to use this design.

At feeding time on each day, each dog is presented with 200 g of Plat du Jour. The data to be analysed is the amount eaten within 10 min. The amounts are shown in Table 18.7.

Table 18.7 Amounts of each recipe eaten

Dog	Day				Mean
	Tuesday	Wednesday	Thursday	Friday	
1	A	B	C	D	
	130	200	175	155	165
2	B	D	A	C	
	180	160	140	200	170
3	C	A	D	B	
	120	105	95	160	120
4	D	C	B	A	
	110	135	190	125	140
Mean	135	150	150	160	149

The mean for each day and for each dog, and the overall mean are given in the margins of the table. The other summary statistics are:

Mean for each recipe:

Recipe	A	B	C	D
Mean	125	182	158	130

18.9   Latin square analysis of variance

The ANOVA for a Latin Square design includes components for the two block variables, 'due to dogs' and 'due to days', and the variable of interest, 'due to recipes'. After their variation is extracted from the total sum of squares the remainder is 'residual'.

The ANOVA table is given in Table 18.8.

Table 18.8   Analysis of variance

Source of variation	Sum of squares	Degrees of freedom	Mean square	Test value
Due to recipes	8525	3	2842	17.95
Due to dogs	6475	3	2158	(13.63)
Due to days	1275	3	425	(2.68)
Residual	950	6	158	
Total	17 225	15		

To determine whether there are significant differences between the acceptability of the recipes, the F-test introduced in Chapter 17 is applied.

Null hypothesis - There is no difference in the acceptability of the four recipes.

Alternative hypothesis - There are differences among the acceptability of the four recipes.

Test value - $= \dfrac{\text{'Due to recipes' mean square}}{\text{Residual mean square}} = \dfrac{2842}{158} = 17.95$

Table value - From Table A.6 (one-sided) with 3 and 6 degrees of freedom: 4.76 at the 5% level, 9.78 at the 1% level.

Decision - Reject the null hypothesis at the 1% level.

Conclusion - There are differences among the recipes in their acceptability.

'Dogs' and 'Days' are the blocks in this experiment. We are not therefore interested in their variation, but it was important to extract their variation from the total variation to leave the true residual that was needed for the test value of the significance test.

The F-ratios for 'due to dogs' and 'due to days' have been shown in parentheses in the ANOVA table so that we can see whether or not the blocking was worthwhile.

The test value for 'due to dogs' was 13.63, compared with test values of 4.76 (5% level) and 9.78 (1% level), shows that there is variability between dogs and therefore it was important to make 'dogs' a block variable. The order effect (due to days) was not significant, but it was still worthwhile including it in the design as a safeguard.

Notice that, in situations where this design would be used for three factors of interest, there is no scope for interactions between the main factors. In the use of a Latin Square design it has to be assumed that interactions do not exist. As the Latin Square is a fractional design, only a limited number of effects can be estimated, and we wish to concentrate on the main effects. In any case, there are insufficient degrees of freedom (only 6 altogether) for interactions.

To complete the analysis, we need the least significant difference between recipe means:

$$5\% \text{ least significant difference} = ts\sqrt{\frac{1}{n_A} + \frac{1}{n_B}}$$

where
  s is the residual SD $= \sqrt{158} = 12.6$ with 6 degrees of freedom,
  t is a coefficient from Table A.2 at the 5% significance level with the same degrees of freedom as s
  and n_A, n_B are the numbers of observations in the formulations being compared.

$$t = 2.45, \ s = 12.6, \ n_A = n_B = 4$$

$$5\% \text{ LSD} = 2.45 \times 12.6\sqrt{\frac{1}{4} + \frac{1}{4}} = 22 \text{ g}$$

Displaying the means, with those that are not significantly different from each other bracketed together:

Recipe	Mean
A	125
D	130
C	157.5
B	182.5

The conclusions are that on the basis of amount eaten, Best Friend's B is the most acceptable, significantly better than each of the others, with a competitor's C the second. Best Friend's A and a competitor's D are not significantly different from each other but significantly less acceptable than the other two.

18.10   Properties and assumptions of the Latin square design

1. The design is chosen so that 3 factors can be evaluated economically.

2. It is appropriate when one of the factors cannot be incorporated into a full factorial design.

3. All factors have the same number of levels.

4. The design is balanced with respect to each pair of factors.

5. Interactions cannot be estimated; they are assumed not to exist.

18.11   Examples of Latin squares

The following are examples of the most commonly used sizes of Latin Square designs. Once the appropriate square has been chosen, the allocation of rows, columns and letters to levels of factors should be randomized.

There are two forms (apart from permutations of rows and columns) of a 3×3 square, several of the 4×4 (some of which are more successful than others at avoiding the repetition of the same pair of treatments) but only the one of the 6×6 square.

3×3

A	B	C		A	B	C
B	C	A		C	A	B
C	A	B		B	C	A

4×4

A	B	C	D		A	B	C	D		A	B	C	D		A	B	C	D
B	A	D	C		B	C	D	A		B	D	A	C		B	A	D	C
C	D	B	A		C	D	A	B		C	A	D	B		C	D	A	B
D	C	A	B		D	A	B	C		D	C	B	A		D	C	B	A

6×6

A	B	C	D	E	F
B	F	D	C	A	E
C	D	E	F	B	A
D	A	F	E	C	B
E	C	A	B	F	D
F	E	B	A	D	C

18.12   Was the experiment successful?

Yes.

The idea of using a Latin Square design was successful in that the dog-to-dog variation did not interfere with comparison of recipes.

If there had been a carry-over effect of any recipe it was balanced in relation to the other recipes.

This experiment had to be carried out over time, so it was worthwhile to include time as another blocking variable. Having ensured balance and blocking, it was possible to detect any significant difference between recipes and conclude that one of the company's own recipes could defeat the competition.

But . . .

For what was essentially a consumer trial the number of 'consumers' was rather small.

A larger group of dogs should have been used and the design could be replicated with each group of four.

Assumptions have been made by humans that the amount of a recipe eaten is a reflection of acceptability. Is there another way of determining it?

18.13   An extra blocking factor – Graeco-Latin square

The Graeco-Latin Square is an experimental design that can investigate four main effects or one main effect and three blocking variables. The levels are balanced in relation to each pair of factors.

The design is so called because of the use of Roman and Greek letters to identify the levels of two of the blocking variables.

One such example of its use might be the study of the effectiveness of four washing powders. The experiment would also use four washing cycles, four machines and be done over four sessions.

An example of Graeco-Latin Square is shown in Table 18.9.

Table 18.9   An example of a Graeco-Latin Square

Machine	Session			
	1	2	3	4
1	$A\alpha$	$B\gamma$	$C\delta$	$D\delta$
2	$B\beta$	$A\delta$	$D\gamma$	$C\alpha$
3	$C\gamma$	$D\alpha$	$A\beta$	$B\delta$
4	$D\delta$	$C\beta$	$B\alpha$	$A\gamma$

The codes in the cells of the table are A, B, C, D for the powders and $\alpha, \beta, \gamma, \delta$ for the washing cycles.

18.14   Problem

The Shimizu Motor Company are investigating several sites in Britain before deciding where to set up their new manufacturing plant. Being mindful of the social welfare of their workers, they have obtained a wealth of information about each location. Part of this has been measurements of the sulfur dioxide levels near each site. The results (ppm) obtained on some preselected days were as follows:

Site	Day				
	Feb. 1	May 1	Aug. 1	Nov. 1	Mean
Kinlochnever	18	10	7	13	12.0
Codurham	41	27	13	31	28.0
Hardunby	34	14	14	34	24.0
Carville	24	17	17	26	21.0
Swestport	20	13	10	17	15.0
Mean	27.4	16.2	12.2	24.2	20.0

(a) Which are the 'blocks' in the experimental design?

(b) Why was it necessary to include a 'block' variable in the design?

(c) Use *Effective* to obtain an analysis of variance table and test whether there is significant variation between sites.

(d) Carry out a multiple range test and decide which sites would be suitable for Shimizu in having the lowest levels.

(e) Would there have been significant differences between sites if 'blocks' had not been incorporated in the design?

19

Two-way designs

19.1 Introduction

In Chapter 17 we met an experiment in which there was one 'controlled' source of variation. The experiment was set up so that a number of 'treatments' could be compared and a decision made about which one was, or which ones were, best.

There is often a need to investigate the effects of two variables, each of which has several choices or can be set at several levels. Examples of typical situations include:

(i) varying washing cycles in several machines to determine their effects on stain removal;

(ii) investigating the effects of time of planting and soil type on the eating quality of varieties of potato;

(iii) the combined effects of pressure and temperature on the yield of a chemical process.

Experiments of this type are factorial experiments, in which the effects of both variables are of interest.

In this chapter we look at an experiment with two factors and replicate results at each set of conditions.

19.2 An example: improving the taste of coffee

Brewfoods are a leading maker of instant coffee. Their Colombian Roast is very highly regarded among customers, but they are constantly seeking to develop and produce a roast that has better taste and in particular to reduce the bitterness that is present.

They have discovered some new origins of beans and wish to compare the roasts from those beans with the origins they already use. The process also involves different roasting methods. As not all beans are expected to respond the same way to different roasting methods, they intend to apply all the roasting methods to all types of beans.

Effective Experimentation: For Scientists and Technologists Richard Boddy and Gordon Smith
© 2010 John Wiley & Sons, Ltd

Once the roasts have been produced they will be assessed by Brewfood's sensory panel.

The Brewfoods bitterness assessment uses a line scale. Each panellist marks on a 9-cm scale his/her assessment of the bitterness of each sample presented. The scores can vary, therefore, from 0 for 'no bitterness' to 9 for 'extremely bitter'. For each sample the mean score is obtained.

The panel assesses three separate samples of each roast. Their assessments are shown in Table 19.1.

Table 19.1 The scores for four bean types and five roasting methods

Origin	Process				
	V	W	X	Y	Z
Jamaica	3.9, 4.1, 4.6	3.5, 3.8, 2.9	4.7, 5.2, 4.8	4.0, 3.3, 3.8	5.9, 5.5, 4.2
Columbia	3.5, 3.2, 3.5	2.5, 2.9, 3.0	3.2, 4.1, 3.8	3.7, 3.0, 2.9	4.2, 3.6, 3.6
Indonesia	5.7, 4.6, 5.0	5.1, 4.6, 5.3	4.2, 3.9, 4.2	4.6, 4.6, 3.7	5.2, 5.2, 4.6
Ethiopia	4.4, 5.1, 5.2	4.6, 4.5, 5.0	4.0, 2.6, 3.0	4.0, 4.9, 3.7	4.3, 3.7, 4.3

Table 19.1 could be described as a 'two-way' table because the scores are categorized in two ways: by **origins** and by **processes**. With four origins and five processes we have 20 **cells** in Table 19.1. As three samples were presented on each occasion we have three scores in each cell and a total of 60 scores within the table.

This two-way classification allows us to speak of both 'origin-to-origin' differences and 'process-to-process' differences.

19.3   Two-way analysis of variance

The statistical method we shall use to analyse the data is analysis of variance.

It will enable us to quantify the variation due to each of these components, 'due to origins' and 'due to processes'.

First, we summarize the data in Table 19.2.

Table 19.2 Mean and standard deviations of the scores in Table 19.1

Origin	Process					Origin means
	V	W	X	Y	Z	
Jamaica	4.2	3.4	4.9	3.7	5.2	4.28
	0.36	0.46	0.26	0.36	0.89	
Columbia	3.4	2.8	3.7	3.2	3.8	3.38
	0.17	0.26	0.46	0.44	0.35	
Indonesia	5.1	5.0	4.1	4.3	5.0	4.70
	0.56	0.36	0.17	0.52	0.35	
Ethiopia	4.9	4.7	3.2	4.2	4.1	4.22
	0.44	0.26	0.72	0.62	0.35	
Process means	4.40	3.98	3.98	3.85	4.52	4.14

The means and standard deviations in Table 19.2 merit careful study by anyone who wishes to get a feel for the variation between processes and between origins. A graph of the means is perhaps even more useful, as we shall see later.

In **two-way** **ANOVA** we can display variation due to origins, based on the variation between the row means, and due to processes, based on variation between the column means. Each of these is obtained in the same way as in Chapters 17 and 18. The residual sum of squares is based on the variation between the three scores in each cell.

The numbers of degrees of freedom are obtained as follows:

Total degrees of freedom:
    = Total number of scores − 1 = 60 − 1 = 59
Due to processes degrees of freedom:
    = Number of processes − 1 = 5 − 1 = 4
Due to origins degrees of freedom:
    = Number of origins − 1 = 4 − 1 = 3
Residual degrees of freedom:
    = (Number of scores in a cell − 1) × (Number of cells) = 2 × 20 = 40

You might have thought that the degrees of freedom (and indeed the sums of squares) of these separate components should, when added, give the total. The sum of the component degrees of freedom is only 4 + 3 + 40, which is 47, whereas the total number of degrees of freedom is 59. This discrepancy has arisen because an essential component of variation is missing from the list. This additional source of variability is known as an **interaction** or, to be more precise, the 'interaction between processes and origins'. It has been included in the ANOVA table (Table 19.3).

Table 19.3 Two-way analysis of variance table

Source of variation	Sum of squares	Degrees of freedom	Mean square
Due to origins	13.76	3	4.586
Due to processes	4.25	4	1.063
Interaction	12.56	12	1.047
Residual	8.20	40	0.205
Total	38.77	59	

We shall see shortly what the meaning of interaction is. We must, however, check for the significance of the interaction before we examine the effects of processes or origins.

Null hypothesis -　　　　　There is no interaction between origins and processes.

Alternative hypothesis -　　There is an interaction between origins and processes.

Test value -　　　　$= \dfrac{\text{Interaction mean square}}{\text{Residual mean square}}$

$= \dfrac{1.047}{0.205} = 5.11$

Table values -   from Table A.6 (one-sided) with 12 and 40 degrees of freedom:
2.00 at the 5% significance level;
2.66 at the 1% significance level.

Decision -   We reject the null hypothesis at the 1% significance level.

Conclusion -   We conclude that there is an interaction between origins and processes.

This test tells us that the interaction cannot be ignored. A graphical representation in Figure 19.1 of the cell means from Table 19.2 will give meaning to the interaction.

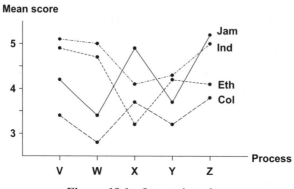

Figure 19.1   Interaction plot.

We can see in Figure 19.1 that Colombia tends to have less bitterness than the other three origins. It is not entirely consistent in this respect, however, for Ethiopia has an even lower score than Colombia with process X. We can also see that Indonesia tends to be more bitter than the others, but that is not entirely consistent across the processes.

The interaction means that the differences between the origins depends upon the process.

Consider the hypothetical graph of three origins that do not interact with processes in Figure 19.2.

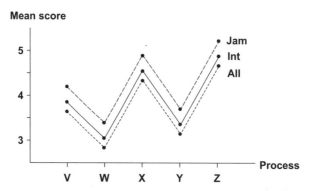

Figure 19.2   No interaction between origins and processes.

In Figure 19.2 are plotted the cell means for Jamaica (from Table 19.2) together with the means of two hypothetical origins Int and All. (This is a hypothetical model; in reality even if there were no interaction there would be some random variation about the parallel lines.) We see that Int receives consistently lower scores than Jamaica, whilst All receives consistently lower scores than Int. Thus, the data in Figure 19.2 contains origin-to-origin variation and process-to-process variation. The parallel plots show that **there is no interaction between origins and processes**.

From Figures 19.1 and 19.2 we can say:

If there is interaction between origins and processes, the differences between origins change from process to process.

If there is no interaction between origins and processes, the differences between origins will be the same with every process.

Let us return to the significance tests on the ANOVA table (Table 19.3). Having found that the interaction mean square is significantly greater than the residual mean square, we must now test the 'due to processes' and the 'due to origins' mean squares against the **interaction**. Having demonstrated the presence of systematic process-to-process variation in the origins (i.e. the interaction) we must now use this as a base line when we check for consistent process-to-process variation and for consistent origin-to-origin variation.

Origins:

Test value - $= \dfrac{\text{'Due to origins' mean square}}{\text{Interaction mean square}}$

$= \dfrac{4.586}{1.047} = 4.38$

Table value - from Table A.6 (one-sided) with 3 and 12 degrees of freedom: 3.49 at the 5% significance level.

Processes:

Test value - $= \dfrac{\text{'Due to processes' mean square}}{\text{Interaction mean square}} = \dfrac{1.063}{1.047} = 1.02$

Table value - from Table A.6 (one-sided) with 4 and 12 degrees of freedom: 3.26 at the 5% significance level.

We can conclude that the 'due to origins' mean square is significantly greater than the interaction whilst the 'due to processes' mean square is not.

We conclude, therefore, that there is a consistent difference between origins over and above the systematic process-to-process variation in origins. There is no overall variation from process to process.

19.4   Multiple comparison test

Now that we have established that there are differences between the origins, we need a least significant difference between the origin means.

Interaction not significant:
If the interaction had not been significant, the calculation would be

$$5\% \text{ LSD} = ts\sqrt{\frac{1}{n_A} + \frac{1}{n_B}}$$

where s is the residual SD,

t from Table A.2 at a 5% significance level has the same degrees of freedom as s and n_A, n_B are the numbers of observations in the two treatments being compared. $s = \sqrt{0.205} = 0.45$ with 40 degrees of freedom, t (5% significance level, 40 degrees of freedom) $= 2.02$.

Each origin had 15 observations, so $n_A = n_B = 15$.
(If we had been comparing the processes, each had 12 observations.)

$$5\% \text{ LSD} = 2.02 \times 0.45 \sqrt{\frac{1}{15} + \frac{1}{15}}$$
$$= 0.33$$

Interaction significant:

Since the interaction was significant, we need to use the interaction mean square in the calculation:

$$5\% \text{ LSD} = t \sqrt{\text{Interaction mean square} \times \left(\frac{1}{n_A} + \frac{1}{n_B}\right)}$$

where t from Table A.2 at a 5% significance level has the same degrees of freedom as the Interaction mean square

and n_A, n_B are the numbers of observations in the two treatments being compared.

Interaction mean square $= 1.047$ with 12 degrees of freedom, t (5% significance level, 12 degrees of freedom) $= 2.18$, $n_A = n_B = 15$.

$$5\% \text{ LSD} = 2.18 \times \sqrt{1.047 \times \left(\frac{1}{15} + \frac{1}{15}\right)}$$
$$= 0.81$$

The significant differences between the origins can be summarized thus:

Origin	Mean
Columbia	3.38
Ethiopia	4.22 ⌐
Jamaica	4.28
Indonesia	4.70 ⌐

We can conclude that the coffee from Columbia is the least bitter, and there are no significant differences among the others.

19.5 Was the experiment successful?

Yes.

Brewfoods have investigated the bitterness of the roasts of coffee from four countries using five roasting methods.

They have established that there is no overall effect of roasting method on the bitterness of the drink.

They have discovered differences between the origins in bitterness.

But ...

There was a significant interaction between origins and processes.

This indicates that although Colombian came out least bitter overall, the choice of best coffee may depend on the roasting method.

The conclusions related only to sensory assessment of bitterness. Other qualities should be investigated.

19.6 Problem

Eatowt are concerned about the coating on their antiulcer drug. Tests are showing a high percentage of broken edges. They believe this could be due to the position of the spray gun and also believe that it could be improved by a change of resin. They carry out an experiment with duplicate tests at each combination of conditions and obtain the following failure percentages:

Position of spray gun	Resin							
	Present		A		B		C	
Top	20.6	25.5	15.0	19.3	12.1	12.8	17.2	14.8
Middle	18.2	15.2	7.4	11.6	7.8	5.1	14.2	19.0
Bottom	25.5	23.9	17.2	18.8	15.1	10.8	14.3	16.4

Draw up an analysis of variance showing the degrees of freedom.

Use *Effective* to obtain the remainder of the analysis of variance table.

Decide on the significance of resins, spray gun positions and their interaction.

Use a least significant difference to decide which spray gun positions and resins are significantly different. This is calculated by:

$$LSD = ts\sqrt{\frac{1}{n_A} + \frac{1}{n_B}}$$

where n_A and n_B are the number of observations in the treatments and s is the residual SD.

Choose the condition that will give the lowest number of broken edges.

20

Too much at once: incomplete block experiments

20.1   Introduction

In Chapter 18 we met the randomized block experiment in which a number of formulations of a product were compared using a panel of subjects. Each subject was treated with every product. The variability between the subjects was taken account of in the design and analysis of the experiment. Similar experiments could apply to a comparison of manufacturing processes, all being done on each of a number of batches of material, or horticultural experiments over a large field that needed to be split into several blocks, within each of which the land should be homogeneous.

A situation may arise in which a batch size is too small to divide for all the processes that are being investigated, or the number of samples is too large for one assessor to evaluate in a session.

How can we reasonably compare the processes or products if we cannot do a complete randomized block experiment?

We can allocate a subset of the processes to work on one batch, or a number of products to be evaluated by an assessor, but as we know there may be batch-to-batch variability and there is very likely to be assessor-to-assessor variation, how can we control the influence of the 'blocks' on the means to allow us unbiased comparisons?

The choice of the subsets to allocate is crucially important.

20.2   Example: an incomplete block experiment

Waxon Products Ltd. is a well established manufacturer of furniture polishes. Jeremy Lewis, a product development chemist, has been assigned to a project developing a spray-on polish with a fragrance that will appeal to the younger user. He realizes the paramount importance of

Effective Experimentation: For Scientists and Technologists   Richard Boddy and Gordon Smith
© 2010 John Wiley & Sons, Ltd

fragrance in furniture polishes and he intends, therefore, to pay particular attention to the assessment of fragrance during household trials.

Six fragrances have been selected for extensive testing in the home, a competitor that is coded D and five Waxon formulations (A, B, C, E and F). Jeremy would like to know if any of his formulations can compete with fragrance D. It is hoped that each user participating in the trial will compare several of these polishes, using each in a separate room of the house. Despite the obvious inconvenience, the use of several rooms is desirable because a person often reacts to a polish fragrance as he or she enters a room.

Dr Lewis realizes that it would be very foolish to present each user with more than three fragrances if he wishes each to be assessed in a different room. On the other hand, it is essential that all six fragrances are evaluated. After careful consideration he decides to use a **balanced incomplete block** experimental design, in which each of ten users is given three tins of spray-on polish, with each tin having a different fragrance. The design is shown in Table 20.1.

Table 20.1 Incomplete block design 6 treatments, 10 blocks, 3 treatments per block

Block (User)	Treatment (Fragrance)					
	A	B	C	D	E	F
1	X	X	X			
2	X	X		X		
3	X		X		X	
4	X			X		X
5	X				X	X
6		X	X			X
7		X		X	X	
8		X			X	X
9			X	X	X	
10			X	X		X

Obviously it would not serve the purpose of the experiment for each user to have the **same** three fragrances. In fact, as we can see in Table 20.1, every user was given a set of fragrances that differed from those given to anyone else.

Notice how this design is balanced – each fragrance is given to the same number of users (5), each user receives the same number of samples (3). In addition to that, each pair of fragrances occurs the same number of times (2 users in common).

Each user was asked to rate each of the three polishes presented to her. Ratings were then converted to numerical scores using 9 = 'Like extremely' ... 1 = 'Dislike extremely'. The scores are given in Table 20.2.

If each of the ten users had rated every one of the six fragrances then Table 20.2 would have contained 60 scores. The experimental design would have been 'complete' and we could have used two-way analysis of variance to estimate a residual standard deviation that could have been used in a comparison of fragrance-means. Unfortunately, it was not possible for every user to rate every fragrance so we have an 'incomplete' experimental design and comparisons will therefore be more difficult.

To make comparisons among the fragrances it would not be fair to compare the means of the data directly, because no two fragrances were assessed by the same set of users.

Table 20.2 Scores given by ten users to three fragrances

User	A	B	C	D	E	F	Mean
			Fragrance				
1	6	5	4				5
2	7	3		5			5
3	4		2		3		3
4	7			3		8	6
5	6				4	8	6
6		2	1			6	3
7		4		4	4		4
8		1			2	6	3
9			1	6	2		3
10			2	2		2	2
Mean	6	3	2	4	3	6	4

For example, there appears to be a big difference between fragrances A and B, but on inspection of the data, users 3, 4 and 5 who assessed A tended to give high scores to other fragrances and users 6, 7 and 8 who assessed B tended to give low scores to other fragrances. As we would expect in consumer testing, there will be biases among the users. Some will tend to give higher scores to everything, some will tend to give lower scores.

The means need to be adjusted to take account of the biases of the users who contributed to each mean, but how do we determine the biases? The means of the users are dependent on the fragrances that they assessed, which we hope will be different!

With such thoughts in mind we could return to Table 20.2 and follow a circuitous path through the 10 users and the 6 fragrances in an attempt to establish fair comparisons.

Fortunately, provided the experimental design is properly balanced there are formulae that allow us to determine the adjustments so that, in effect, we are predicting what the means would have been from the full panel.

20.3   Adjusted means

The adjustment for each fragrance mean is given as follows:

Adjusted fragrance mean

$$= \text{Observed fragrance mean}$$
$$- \left(\frac{fb - N}{f(N-b)} \right) \times (\text{Sum of other fragrance means})$$
$$+ \left(\frac{f-1}{N-b} \right) \times (\text{Sum of other user means})$$

where $f =$ number of treatments (fragrances)
     $b =$ number of blocks (users)
and $N =$ total number of scores.

In this experiment Jeremy has compared **six** fragrances using **ten** users who between them awarded **thirty** scores. Substituting $f = 6$, $b = 10$ and $N = 30$ into the formula gives us:

Adjusted fragrance mean = Observed fragrance mean $- 0.25$ (Sum of other fragrance means) $+ 0.25$ (Sum of other user means)

To obtain the adjusted mean for fragrance A we proceed as follows:

$$\text{Mean for fragrance } A = 6.0$$

$$
\begin{aligned}
\text{Sum of other fragrance means} &= \text{Sum of means for } B, \ C, \ D, \ E \text{ and } F \\
&= 3.0 + 2.0 + 4.0 + 3.0 + 6.0 \\
&= 18.0
\end{aligned}
$$

$$
\begin{aligned}
\text{Sum of other user means} &= \text{Sum of means for all users who did } \textbf{not} \text{ assess } A \\
&= \text{Sum of means for } 6, \ 7, \ 8, \ 9 \text{ and } 10 \\
&= 3.0 + 4.0 + 3.0 + 3.0 + 2.0 \\
&= 15.0
\end{aligned}
$$

$$
\begin{aligned}
\text{Adjusted mean for fragrance } A &= 6.0 - 0.25(18.0) + 0.25(15.0) \\
&= 5.25
\end{aligned}
$$

So, as we suspected, fragrance A was assessed by users who tended to score high, and once the mean was adjusted to take account of their biases, the mean became 5.25. This is what we predict would have been given if the whole panel had assessed it.

The means are adjusted as shown in Table 20.3.

Table 20.3 Calculation of adjusted fragrance-means

Fragrance	A	B	C	D	E	F
Observed fragrance mean	6.0	3.0	2.0	4.0	3.0	6.0
Sum of other fragrance means	18.0	21.0	22.0	20.0	21.0	18.0
Sum of other user means	15.0	20.0	24.0	20.0	21.0	20.0
Adjusted fragrance means	5.25	2.75	2.50	4.00	3.00	6.50

The adjusted means in Table 20.3 are much more useful than the observed means on which they are based. It can be shown that **the adjusted fragrance-means are unbiased estimates of the means we would have obtained if every user had rated every fragrance**. The unadjusted means, on the other hand, are biased by the generosity or meanness of the users to which the fragrance was assigned.

We can also see now that fragrances C and F suffered from low-scoring users, especially user 10 who disliked all his three fragrances so much he never scored above 2. Their means are adjusted upwards.

20.4 Analysis of variance for balanced incomplete block design

As the prime objective of Jeremy's experiment is to compare the six fragrances and in particular to see if the competition can be beaten, he would wish to make use of the adjusted fragrance means in Table 20.3. He is particularly interested to know if fragrance F with the highest adjusted mean (6.50) is significantly more liked than fragrance D, the competitor's brand.

If all users had assessed all fragrances, we would have been able to analyse the data using the ANOVA that we used for the randomized block experiment in Chapter 18. We can still use a form of ANOVA, but it needs to be modified for the incomplete design. The ANOVA table is shown in Table 20.4.

Table 20.4 Analysis of variance for balanced incomplete block experiment

Source of variation	Sum of squares	Degrees of freedom	Mean square
Due to fragrances (adjusted)	50.5	5	10.1
Due to users	54.0	9	Not to be used
Residual	25.5	15	1.70
Total	130.0	29	-

The sums of squares for 'total' and 'due to users' are calculated in the same way as in Chapter 18.

$$\text{Adjusted between fragrances sum of squares}$$
$$= (N{-}b)\,(\text{SD of adjusted fragrance means})^2$$

where N is the total number of scores, b the number of blocks (users) and the SD of adjusted fragrance means, obtained from the means in Table 20.3, is 1.5890.

$$\text{Adjusted between fragrances sum of squares}$$
$$= (30{-}10)\,(1.5890)^2$$
$$= 50.5$$

Let us check the degrees of freedom:

$$\text{Total degrees of freedom} = \text{total number of scores}{-}1 = 29$$
$$\text{Due to fragrances : number of fragrances}{-}1 = 5$$
$$\text{Due to users : number of users}{-}1 = 9$$
$$\text{Residual : by difference} = 29{-}5{-}9 = 15$$

We can use the F-test to determine whether there are any differences between the fragrances in terms of their popularity.

Null hypothesis -   There is no difference in the popularity of the six fragrances.

Alternative hypothesis -   There are differences in the popularity of the six fragrances.

Test value - $= \dfrac{\text{'Due to fragrances' mean square}}{\text{Residual mean square}}$

$$= \frac{10.1}{1.70} = 5.94$$

Table value -   From Table A.6 (one-sided) with 5 and 15 degrees of freedom: 2.90 at the 5% level, 4.56 at the 1% level.

Decision -   Reject the null hypothesis at the 1% level.

Conclusion -   There are differences in the popularity of the six fragrances.

Thus, we can confidently conclude that the fragrances do differ in their appeal to the users. We now wish to discover where the differences are.

Before we can calculate a least significant difference, we need the residual standard deviation.

$$\text{Residual standard deviation} = \sqrt{\text{Residual mean square}}$$

$$= \sqrt{1.70} = 1.304 \text{ with 15 degrees of freedom}$$

The formula for a 5% least significant difference between adjusted fragrance means is an amended version of the one that we used in the randomized block experiment in Chapter 18:

$$5\% \text{ least significant difference} = t \times (\text{Residual SD}) \times \sqrt{\frac{2(f-1)}{N-b}}$$

where f = number of treatments (fragrances)

$\quad\quad b$ = number of blocks (users)

$\quad\quad N$ = total number of scores

and t is a value from Table A.2 at a 5% significance level with the same degrees of freedom as the residual SD.

$$f = 6, \ b = 10, \ N = 30, \ t = 2.13 \text{ with 15 degrees of freedom}$$

$$5\% \text{ LSD} = 2.13 \times 1.304 \times \sqrt{\frac{2 \times 5}{30-10}} = 1.96$$

If two adjusted fragrance means differ by more than 1.96 we can be 95% confident that the fragrances are not equally well liked.

We can display the adjusted means, in ascending order, in Table 20.5, bracketing together those that are not significantly different from each other.

Table 20.5 Significant differences between the six fragrances

Fragrance	Adjusted mean
C	2.50
B	2.75
E	3.00
D	4.00
A	5.25
F	6.50

From Table 20.5 we can conclude that fragrance F is significantly preferred to the competitors' D. F is significantly preferred to all the fragrances except A, while all the others – A, B, C, and E are not significantly more or less acceptable than D.

20.5   Alternative designs

The design that Jeremy used in the above example is appropriate for six treatments, 10 blocks and 3 treatments per block. There are other designs of other sizes, for example, the design shown in Table 20.6 is for 5 treatments, 10 blocks and 3 treatments per block.

Table 20.6   Incomplete block design 5 treatments, 10 blocks, 3 treatments per block

Block	Treatment				
	A	B	C	D	E
1	X	X	X		
2	X	X		X	
3	X	X			X
4	X		X	X	
5	X		X		X
6	X			X	X
7		X	X	X	
8		X	X		X
9		X		X	X
10			X	X	X

20.6   A design with a control

It is sometimes desirous to compare treatments with a control that will occur in every block. For the same total number of observations, the number of occurrences of the other treatments would be reduced, and hence the power of comparisons among them would be less, but if the main objective is comparison with the control, this would not matter.

A suitable design is set out in Table 20.7.

Table 20.7   An incomplete experiment with a control 6 treatments, 10 blocks, 30 observations

Block	Treatment					
	A	B	C	D	E	F
1	X	X	X			
2	X		X	X		
3	X		X		X	
4	X		X			X
5		X	X	X		
6		X	X		X	
7		X	X			X
8			X	X	X	
9			X	X		X
10			X		X	X

The experiment in Table 20.7 is in some ways more appealing than the balanced incomplete block experiment in Table 20.1. As the two experiments share the same objective, to compare the six fragrances, let us see what comparisons could have been made if Jeremy had persisted with the design in Table 20.7. If we wish to compare A and C we find that the scores of 4 users offer a direct comparison. This would be true if we compare any fragrance with the control fragrance, C. If on the other hand we wish to compare fragrances A and B we find that only one user rated these two fragrances so the direct comparison is much less precise. Both of the experimental designs also offer a host of indirect comparisons but these are not so easy to evaluate.

If, therefore, we wish to make equally precise comparisons of any two fragrances the balanced incomplete block experiment of Table 20.1 is superior to the design in Table 20.7. On the other hand, if one particular fragrance has special significance then designating it as a control will improve the comparison of this fragrance with any other. Unfortunately the price we must pay for this improvement is a decrease in precision with which other pairs of fragrances can be compared.

20.7   Was Jeremy's experiment successful?

Yes.

He used a suitable experimental design to allow him to compare the acceptabilities of six fragrances when it was practicable for each user to assess only three.

The design and analysis enabled him to obtain mean scores that were adjusted for user biases.

He was able to conclude that his formulations could stand competition with that of a competitor, and one of them was significantly preferred to the competitor's.

But...

If one of the purposes was to compare all his formulations with the competitor's, should he not have used the design in which the control was assessed by every user?

For this type of consumer trial, he should have had a larger panel.

His immediate intention is to extend this investigation to include a further 20 users. He will carry out two additional replicates of what he has already completed. Jeremy is well aware that the validity of his conclusions depends upon the representativeness of his sample of users.

20.8   Problem

Dr. Fox of Glacier Mints Ltd. wishes to compare 5 experimental flavours that have been developed for use in a new style of chocolate mint. His experience in the sensory evaluation of chocolates suggests that assessors should not be asked to rate more than three samples in one session. He therefore decides to use a balanced incomplete block experiment involving 10 assessors. The acceptability scores given to the 5 flavours by the assessors are in the table below.

Assessor	Flavour					Mean
	A	B	C	D	E	
1	7	3	5			5
2	6	2		1		3
3	7	3			5	5
4	5		5	2		4
5	7		7		7	7
6	4			2	6	4
7		6	8	1		5
8		6	9		6	7
9		4		3	5	4
10			8	3	7	6
Mean	6		7	2	6	5

(a) Calculate an adjusted mean for each of the five flavours.

(b) Produce an analysis of variance table using the adjusted flavour means and the crude assessor means.

(c) Are there any differences among the flavours?

(d) From the analysis of variance table calculate an estimate of the residual standard deviation.

(e) Calculate the least significant difference and decide which pairs of flavours are significantly different.

21

23 Ways of messing up an experiment

21.1   Introduction

In this chapter we first look at the ways of messing up an experiment, many of which will be familiar to us from our own experience. Then, building upon this knowledge, we propose a sequence of necessary considerations in the planning, design, execution and analysis of an experiment.

However, before we start it is perhaps as well to bear in mind an important truism:

'*The best time to design an experiment is after it is finished . . . .*
the worst time is at the beginning, when least is known'.

We therefore must expect to make some errors of judgement because of our lack of knowledge of the process under consideration. We should, however, try to avoid mistakes which are caused by our lack of knowledge of experimental design.

21.2   23 ways of messing up an experiment

(1) Attacking one response at a time.

(2) Attacking one variable at a time.

(3) Ignoring the possibility of interactions.

(4) Exploring too small a range of the variables.

(5) Designing always to obtain good responses. Poor responses - but not too bad - give valuable information.

(6) Failing to recognize and deal with nuisance variables.

(7) Failing to deal with important process variables which are not under investigation.

(8) Assuming all the levels will run and give reasonable responses.

(9) Putting too much emphasis on curvature (belief in complexity rather than simplicity).

(10) Designing the experiment to optimize the process when there is no knowledge about which variables are important.

(11) Failing to sequence trials appropriately; this could result in a time trend in the response looking like a significant effect.

(12) Over-designing the experiment so that too many variables are investigated in too few trials, thus not allowing any check on the validity of conclusions.

(13) Believing you are the seat of all wisdom and that you therefore can achieve the optimum in one trial - use a systematic approach - it may be boring but it will succeed in the end.

(14) Too much fiddling with the levels of the variables. Do you really need a temperature level of $289°C$ when you already have $288°$ and $290°$?

(15) Not consulting the operational staff before the experiment.

(16) Operators unable to understand instructions.

(17) Operators unable to follow instructions.

(18) Operators unwilling to follow all instructions.

(19) Continuing with the experiment when it is clearly going wrong.

(20) Not admitting the possibility of outliers in the data.

(21) Throwing out observations, on the excuse that they are outliers, because they don't fit your theory.

(22) Failing to plot the data and search for abnormal patterns.

(23) Overfitting the data by including too many variables and interactions so that you are fitting parameters to the errors rather than to the underlying effects.

21.3   Initial thoughts when planning an experiment

(a) What are the objectives?

These should be expressed in precise terms. For example:-

'We wish to determine that there is no adverse effect on the quality of a new cheaper catalyst PB212. We will consider PB212 to have an adverse effect if it results in whiteness declining by 1% or opaqueness by 2%'.

(b) Have you examined the literature or any previous experience?

Previous experience will be extremely valuable to the design but we should always check that it is relevant and not based upon conditions which no longer apply.

(c) Have you consulted the personnel who will run the experiment? They must have much knowledge of the process as they run it full-time.

21.4   Developing the ideas

(a) What type of experiment is required?

  (i) An experiment to investigate the importance of the variables with a view to improving some property of the product?

  (ii) An optimization experiment in which knowledge about the independent variables is not the major consideration?

  (iii) A ruggedness experiment in which the importance of changes in levels of independent variables is investigated with a view to ensuring that the product quality is capable of conforming to a specification?

(b) What are the independent variables?

(c) What range should be investigated for each independent variable?

(d) What interactions may be important?

(e) What are the sources of error and what is their likely magnitude?

Possible sources are:-

Sampling error   Testing bias
Testing error (or replicates) Batch-to-batch error (raw materials, time, etc.)

21.5   Designing the experiment

(a) Are the independent variables continuous variables (temperature, time) or qualitative variables (machine, treatment)? If they are qualitative variables, how many treatments need to be investigated?

(b) What should be done with nuisance variables such as a variable feedstock, important process variables which are of no interest to the investigation and time factors? The methods of dealing with these, in order of preference, are given below:-

  (i) Control them so they don't vary.

  (ii) Balance them, in blocks, so that their effect can be removed and not cause bias in the stipulated effects.

  (iii) Measure them and remove their effect, if necessary, using multiple regression.

  (iv) Ensure that they are randomized over all the experiments so that they don't cause any bias.

(c) What responses will be measured?

(d) How many independent variables should be included in the design?

(e) Can the investigation be divided into several small experiments to guide you in the right direction? This will allow you to update the design after each one. The division will need to be based on the requirement that variables comprising important interactions must be included in the same experiment.

(f) How many trials are required to give a reasonable chance of establishing that effects are significant?

(g) Is a sighting experiment (to fix our sights as used in rifle shooting) required to check that the majority of trials will run and produce responses of some interest? Trials are to be avoided if they all give similar responses or if they contain many extreme values of no interest to future production.

(h) What sequence of trials should be used in order to minimize the residual SD and to prevent bias?

21.6   Conducting the experiment

(a) Is the plan too complicated for the operators to run easily?

(b) Should the order of trials be changed so that less demands are made on the operators? The sequence of trials could be arranged so that a 'difficult' or 'expensive-to-change' variable is only altered a few times during the investigation.

(c) What procedure will be adopted if a trial does not run or gives rogue values? Should another trial at different levels be substituted? Should an attempt be made to repeat the trial? Should the experiment be stopped?

21.7   Analysing the data

(a) What plotting procedures should be attempted?

  (i)   Residuals versus independent variables?

  (ii) Normal plot of residuals?

  (iii) Residuals or contour diagrams?

  (iv) Half-normal plot?

(b) Are there any possible outliers in the data?

(c) Do the significant effects and coefficients make technical sense? For example, have you obtained a negative effect for a variable which must have a positive effect?

(d) Have you 'removed' the effect of nuisance variables from the residual SD?

(e) Is the residual SD similar to those obtained in previous experiments or does it relate sensibly to your experience of the process?

21.8   Summary

The planning and designing of an experiment are at least as important as its execution and the analysis of the results. Good planning and design requires some knowledge of the principles and techniques of experimental design but above all it requires careful thought and detailed preparation. There can be no substitute for good design; problems not dealt with at the design stage (or worse, problems built into the design) can never be fully removed later, and could lead to false or indeterminate conclusions.

Solutions to problems

Chapter 2

a)

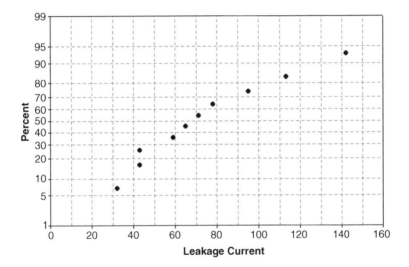

The normal probability plot shows that the data is not a good fit to a straight line but equally the distribution is clearly not very skewed or the fit would be poorer. Fortunately the two-sample t-test is not very dependent upon the assumption of normality. Providing there is a reasonable sample size - and 10 and 8 are very reasonable - and the data is not highly skewed, the test will be valid.

Effective Experimentation: For Scientists and Technologists   Richard Boddy and Gordon Smith
© 2010 John Wiley & Sons, Ltd

b) 95% confidence interval for the population mean is given by:

$$\bar{x} \pm \frac{ts}{\sqrt{n}}$$

For the standard process, $\bar{x} = 74.1$, $t = 2.26$ (from Table A.2 with 9 degrees of freedom at the 95% confidence level), $s = 34.32$, $n = 10$.

$$74.1 \pm \frac{2.26 \times 34.32}{\sqrt{10}} = 74.1 \pm 24.5$$

i.e. 49.6 to 98.6

For the modified process, $\bar{x} = 67.0$, $t = 2.31$ with 7 degrees of freedom, $s = 32.15$, $n = 8$.

$$67.0 \pm \frac{2.31 \times 32.15}{\sqrt{8}} = 67.0 \pm 26.9$$

i.e. 40.1 to 93.9

There is considerable overlap between the two confidence intervals so it is doubtful whether they will be significantly different. However, if the overlap was small it would be possible.

c)

Null hypothesis - The standard and modified processes have the same population means. ($\mu_A = \mu_B$)

Alternative hypothesis - The population means are **not** equal. ($\mu_A \neq \mu_B$)

Test value - Where A = standard process and B = modified process

$$Combined\ SD(s) = \sqrt{\frac{(df_A \times SD_A^2) + (df_B \times SD_B^2)}{df_A + df_B}}$$

$$= \sqrt{\frac{(9 \times 34.32^2) + (7 \times 32.15^2)}{9 + 7}}$$

$$= 33.39 \text{ with } 16 \text{ degrees of freedom}$$

$$Test\ value = \frac{|\bar{x}_A - \bar{x}_B|}{s\sqrt{\frac{1}{n_A} + \frac{1}{n_B}}}$$

$$= \frac{|74.1 - 67.0|}{33.39\sqrt{\frac{1}{10} + \frac{1}{8}}} = 0.45$$

Table value - From Table A.2 with 16 degrees of freedom: 2.12 at the 5% significance level.

Decision - We cannot reject the null hypothesis.

Conclusion - With this evidence, Seltronics would be unwise to change to the modified process.

Chapter 3

a)

Customer	A	B	C	D	E	F	G	H	Mean	SD
1st Quarter	68	82	87	72	80	83	94	66	79.0	9.67
2nd Quarter	63	75	85	73	71	78	88	59	74.0	9.93
Difference (1st - 2nd)	5	7	2	−1	9	5	6	7	5.0	3.16

For each customer we have subtracted the Quarter 2 result from the Quarter 1 result to form a difference. We therefore now have a sample of 8 differences from a population of differences and we can use a one-sample t-test to see if the mean difference is not equal to zero. In examining the table above we notice that the standard deviation of the differences is greatly reduced from that of the original results, thus vindicating the decision to use a matched-pair design.

We now proceed to use the paired t-test, which is just a one-sample t-test on the differences.

Null hypothesis -   The population means for Quarter 1 and Quarter 2 are equal, i.e. the mean of the population of differences is equal to zero. ($\mu_d = 0$)

Alternative hypothesis -   The mean of the population of differences is **not** equal to zero. ($\mu_d \neq 0$)

Test value - $\dfrac{|x_d - \mu_d|\sqrt{n_d}}{s_d} = \dfrac{|5 - 0|\sqrt{8}}{3.16} = 4.47$

Table values -   From Table A.2 with 7 degrees of freedom: 2.36 at the 5% significance level.

Decision -   Since the test value is larger than the table value we can conclude there is a significant difference between quarters.
There is no doubt that the mean score has reduced and it could be concluded that the company's performance had deteriorated.

b) Sample size

$$n_d = \left(\frac{ts_d}{c}\right)^2$$

$$= \left(\frac{2.36 \times 3.16}{3}\right)^2 = 6.2$$

Thus a sample size of 7 differences is required. This is the same 7 retailers having scores in both quarters.

c) The combined SD would be:

$$Combined\ SD(s) = \sqrt{\frac{(df_A \times SD_A^2) + (df_B \times SD_B^2)}{df_A + df_B}}$$

$$= \sqrt{\frac{(7 \times 9.67^2) + (7 \times 9.93^2)}{7 + 7}}$$

$$= 9.80 \text{ with 14 degrees of freedom}$$

$$Sample\ size\ (= n_A = n_B) = 2\left(\frac{ts}{c}\right)^2$$

where t = 2.15 with 14 degrees of freedom at the 5% significance level

$$= 2\left(\frac{2.15 \times 9.80}{3.0}\right)^2 = 99$$

Thus 99 retailers would be needed in each quarter. This shows the great benefit of paired designs - the sample size is 99 for independent samples compared with 7 for paired samples. Although pairing can not often be applied, where it can it yields considerable benefits. Thus the experimenter should always be looking out for opportunities to apply it.

Chapter 4

1. (a) Two variables have been used.
   (b) Each variable has two levels.
   (c) Each cell is duplicated.
   (d) A $2(2^2)$ factorial experiment.
   (e)

Cement Resin	10%	12%
1.0%	$\bar{x} = 530$ s = 8.49	$\bar{x} = 550$ s = 2.83
1.5%	$\bar{x} = 542$ s = 5.66	$\bar{x} = 535$ s = 7.07

(f) Main effect of cement $= \frac{1}{2}(550 + 535) - \frac{1}{2}(530 + 542)$ $=$ 6.5

Main effect of resin $= \frac{1}{2}(542 + 535) - \frac{1}{2}(530 + 550)$ $= -1.5$

Interaction of cement and resin $= \frac{1}{2}(530 + 535) - \frac{1}{2}(542 + 550) = -13.5$

(g) Residual standard deviation

$$= \sqrt{\frac{(8.49)^2 + (2.83)^2 + (5.66)^2 + (7.07)^2}{4}}$$

$= 6.36$ based on 4 degrees of freedom

(h) The 95% confidence intervals are given by $\pm \dfrac{2ts}{\sqrt{n}}$

$s = 6.36$; t (Table A.2, with 4 degrees of freedom at 95% confidence level) $= 2.78$; $n = 8$

Confidence interval is

$$\pm \frac{2 \times 2.78 \times 6.36}{\sqrt{8}} = \pm 12.5$$

Thus the interaction (-13.5 ± 12.5) is significant but the main effects are not.

(i) On inspecting the table of means, the conditions which give the grout with the highest crushing strength are 1.0% resin, 12% cement which would give a predicted mean crushing strength of 550. This is only the best of the four conditions tested in the experiment. Some more experimentation should be done in the region of this combination to confirm the conclusion.

2. (a) Combined estimate of Residual Standard Deviation (RSD) is:

$$\sqrt{\frac{8.5^2 + 8.5^2 + 5.7^2 + 5.7^2}{4}}$$

$= 7.2$ based on 4 degrees of freedom

Estimating main effects:-

Pressure Temperature	1.3	1.6	Mean
271	219	262	240.5
279	251	285	268
Mean	235	273.5	

256.5

252.0

95% confidence interval for each effect

$$= \pm \frac{2ts}{\sqrt{n}}$$

$$= \pm \frac{2 \times 2.78 \times 7.2}{\sqrt{8}}$$

$$= \pm 14.2$$

Variable	Effect
Pressure	38.5
Temperature	27.5
P × T	−4.5

Both Pressure and Temperature have significant positive effects but the interaction is not significant.

(b) Notice that the two low-speed conditions (50 r.p.m.) coincide with one diagonal in the Pressure-Temperature 2 × 2 table and the high-speed conditions (60 r.p.m.) coincide with the other diagonal. Thus the difference of −4.5 between the diagonal means could be due to an effect of Speed or could be some combination of a Speed effect and a P × T interaction. Another experiment would be necessary to sort out this confusion.

Chapter 5

1. (a) Results in standard order:

Trial	A	B	AB	C	AC	BC	ABC	Brightness
6	−	−	+	−	+	+	−	7
1	+	−	−	−	−	+	+	5
3	−	+	−	−	+	−	+	5
7	+	+	+	−	−	−	−	1
5	−	−	+	+	−	−	+	3
4	+	−	−	+	+	−	−	9
8	−	+	−	+	−	+	−	3
2	+	+	+	+	+	+	+	7

(b) Main effect of speed (A) $= (-7 + 5 - 5 + 1 - 3 + 9 - 3 + 7)/4 = 1.0$
Similarly B $= -2.0$; AB $= -1.0$; C $= 1.0$; AC $= 4.0$; BC $= 1.0$; ABC $= 0.0$.

(c)

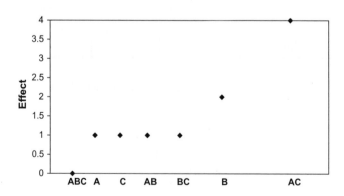

The half-normal plot suggests that 5 effects should be used for the Residual SD.

Residual SD $= 1.26$ with 5 degrees of freedom

95% confidence interval for effect : ± 2.3

Thus only the interaction AC (4.0) is significant.

(d) The 95% confidence interval for effect estimate

$$= \pm \frac{2ts}{\sqrt{n}}$$

(t with 3 degrees of freedom is 3.18 at the 95% confidence level)

$$= \frac{2 \times 3.18 \times 0.80}{\sqrt{8}}$$

$$= \pm 1.80$$

The main effect B and interaction AC are significant by this method.

The significance of AC is confirmed; there is perhaps some doubt about the significance of B.

2. (a) The effects are:-

A	1.38	D	-1.53
B	-0.62	AD	-0.08
AB	-0.28	BD	0.02
C	0.87	ABD	0.48
AC	3.52	CD	0.12
BC	0.52	ACD	0.67
ABC	0.38	BCD	0.18
		ABCD	-0.27

(b)

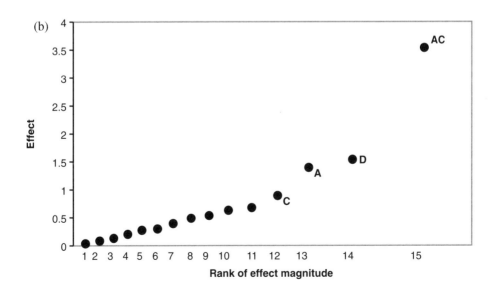

We see from the half-normal plot that there is little doubt about the significance of AC, but we are less certain about A and D. Although C looks less likely to be significant, in view of its inclusion in the important interaction AC we decide to exclude it from the residual.

$$\text{With 11 effects, Residual SD} = 0.78$$
$$95\% \text{ confidence interval for effect estimates} = \pm 0.86$$

Thus AC, A, C and D are all significant although there is little doubt that AC dominates the relationship with brightness.

(c) The significance of AC is confirmed; the doubts about B were well-founded, and with a larger set of data A and C are now significant.

3. (a) Rearranging the data in standard design matrix form gives:-

P	T	S	PTS	Plasticity	Experiment
−	−	−	−	219	1
+	−	−	+	281	2
−	+	−	+	268	2
+	+	−	−	285	1
−	−	+	+	228	2
+	−	+	−	262	1
−	+	+	−	251	1
+	+	+	+	283	2

Combined RSD = 7.93 (8 degrees of freedom)

Analysis:-

Variable	Effect
P	36.25
T	24.25
S	−7.25
PT	−11.75
PS	−3.25
TS	−2.25
PTS	10.75

(b) 95% confidence interval

$$= \pm \frac{2ts}{\sqrt{n}}$$

(t with 8 degrees of freedom is 2.31 at the 95% confidence level)

$$= \frac{2 \times 2.31 \times 7.93}{\sqrt{16}}$$

$$= \pm 9.2$$

(c) Interpretation of Effects:

The temperature and pressure effects are both highly significant and positive, suggesting a move towards higher values for both these variables.

The pressure x temperature interaction is also significant at the 5% level and negative, implying that the effect of temperature is decreasing with higher pressure (or that the effect of pressure is decreasing with higher temperature). If this same interaction effect continues to the higher levels of pressure and temperature the highest plasticity might not be at the highest possible values of temperature and pressure. It is always worthwhile looking at the appropriate 2 × 2 table, given below, when an interaction is significant, as this shows clearly what is happening.

	P	
T	1.3	1.6
271	223.5	271.5
279	259.5	284.0

The table indicates that the gain from increasing temperature is less at a pressure of 1.6. This is unfortunate since we are searching for high plasticities.

The effect of speed is not significant, but it has a value not too much less than the 5% least significant difference and is negative, suggesting that it would be dangerous to ignore the possibility that over the wider range of speeds possible it might have an effect.

The PTS interaction is also significant at the 5% level. It is fairly unusual to find three-variable interactions significant and it is worth looking to see if any other explanation is available. Studying the signs of the PTS interaction column in the design matrix it is noticed that all the negative signs coincide with the first set of experiments and the positive signs with the second set. Thus this interaction could also be explained as a difference in plasticity level between the two sets of experiments.

Chapter 6

1. The variables are coded as follows:-

Recycle feed rate (A), Pressure (B), Evaporation level (C), Steam rate (D). The stipulated effects are A, B, C, D, AB, BC and BD.

Effect	Alias
A*	BCD
B*	ACD
AB*	CD
C*	ABD
AC	BD*
BC*	AD
ABC	D*

We notice that by using ABCD as the defining contrast, none of the alias pairs contains two stipulated effects. So ABCD is a suitable defining contrast, but it is not the only one since ACD also gives a suitable pattern of alias pairs.

Trial	A	B	C	D = ABC
1	−	−	−	−
2	+	−	−	+
3	−	+	−	+
4	+	+	−	−
5	−	−	+	+
6	+	−	+	−
7	−	+	+	−
8	+	+	+	+

We can now translate the design matrix to give the conditions for the four variables:-

Trial	Recycle Feed Rate (A)	Pressure (B)	Evaporation Level (C)	Steam Rate (D)
1	Low	Low	Low	Low
2	High	Low	Low	High
3	Low	High	Low	High
4	High	High	Low	Low
5	Low	Low	High	High
6	High	Low	High	Low
7	Low	High	High	Low
8	High	High	High	High

2.

A (or BCD) = 3
B (or ACD) = 13.5
AB (or CD) = −3
C (or ABD) = −0.5
BD (or AC) = 17
BC (or AD) = −2.5
D (or ABC) = 39

95% confidence interval for each effect

$$= \pm \frac{2ts}{\sqrt{n}} = \pm \frac{2 \times 2.36 \times 4.1}{\sqrt{8}} = \pm 6.8$$

We can conclude that the Pressure × Steam rate interaction together with the main effects of these variables are significant at the 5% level. Since all the effects are positive the conditions which will maximise the refractive index are high Pressure and high Steam Rate.

The mean results for each level of Pressure and Steam Rate are as follows.

Pressure (B)	Steam Rate (D)	
	Low	High
Low	87	109
High	84	140

This conclusion is valid as long as the researcher has been correct in his assumptions about interactions. He stipulated the BD interaction and assumed that an AC interaction did not exist.

Chapter 7

1. (a) The estimated effects are:-

Effect	Estimate
Effect variables:	
A	0.10
B	−0.30
C	−3.03
D	−0.27
E	2.07
F	0.40
Residual columns:	
G	−0.27
H	0.27
I	0.17
J	0.53
K	0.47

Basing the RSD on the 5 residual columns

$$\text{RSD} = \sqrt{\frac{n}{4} \times \frac{\sum squared\ differences}{no.\ of\ residual\ columns}}$$

$$= \sqrt{\frac{12}{4} \times \frac{0.27^2 + 0.27^2 + 0.17^2 + 0.53^2 + 0.47^2}{5}}$$

$$= 0.64 \text{ with 5 degrees of freedom}$$

The 95% confidence interval for each effect is

$$\pm \frac{2ts}{\sqrt{n}} = \pm \frac{2 \times 2.57 \times 0.64}{\sqrt{12}} = \pm 0.94$$

The significant effects are : Bottom Air Temperature (C) : -3.03 ± 0.94

Baffle Angle (E) : 2.07 ± 0.94

(b)

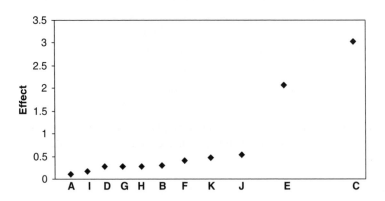

There are nine non-significant effects.

(c) Nine variables should be used in estimating an improved RSD.
This gives an estimate of the RSD of 0.58 with 9 degrees of freedom.
The 95% confidence interval for each effect is

$$\pm = \frac{2ts}{\sqrt{n}} = \pm \frac{2 \times 2.26 \times 0.58}{\sqrt{12}} = \pm 0.76$$

The confidence interval is narrower because the t value is lower with more degrees of freedom, and also because the estimated RSD happened to be lower.

2. The non-significant effects on the Half-Normal plot do not give a line going through zero, which indicates the presence of an outlier.

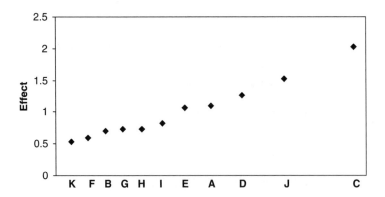

The estimated effects are:-

Effect	Estimate
Effect variables:	
A	1.10
B	0.70
C	−2.03
D	−1.27
E	1.07
F	−0.60
Residual columns:	
G	0.73
H	−0.73
I	−0.83
J	1.53
K	−0.53

The RSD from the residual columns is swollen from 0.64 to 1.63.

Chapter 8

Looking at the statistical analysis, the following observations can be made:

(i) The slopes and the residual standard deviations are all very similar showing the experiments have been carried out with the same precision.

(ii) Cooper has by far the smallest confidence interval for the slope and Dawson has by far the worst. This pattern is also the same for the correlation coefficient.

Why is Dawson's design so poor in terms of the confidence interval for the slope? It is because he has used a very narrow set of conditions for the independent variable - wind-up speed. In designing experiments it pays to be bold. However, this must be tempered that the conditions are realistic. Thus, if a wind-up speed of 150 produced nonsensical results, Cooper's experiment would have been a complete failure. However, using 150 was a suitable choice and thus Cooper's design was the best. He has, however, taken the biggest risk by only using two values and therefore having little chance to evaluate curvature. Thus, if one of his extreme values had been unrealistic or curvature had been present, Cooper would have been a failure.

What about Bolam's design? Having points in the centre enables one to evaluate curvature but having four repeats is unwise since he has only two effective points to establish the slope. If one of these were an outlier his design would have failed

Addy's design is a traditional method of tackling the problem but is far from good. Having many different levels is unnecessary as long as the work is not exploratory.

A better design than any of the four would be three levels and two replications at each level.

Chapter 9

1. The inter-correlation matrix shows that there is going to be trouble ahead.

Population (B)	1.00					
Salespersons (C)	0.93	1.00				
GDP (D)	0.97	0.94	1.00			
Complaints (E)	0.86	0.72	0.83	1.00		
Competitors (F)	0.20	0.16	0.26	0.53	1.00	
Advertising (G)	0.72	0.45	0.60	0.73	0.04	1.00
	Pop. (B)	**Salesp. (C)**	**GDP (D)**	**Compl. (E)**	**Compet. (F)**	**Advert. (G)**

With so many high inter-correlations there are many alternative models, e.g.

$$Sales = 119 + 80B - 2.2E - 34.0F + 25.5G \quad \%fit = 98.3$$
$$Sales = 331 + 151B - 11.1C - 83.5F \quad \%fit = 92.8$$
$$Sales = 273 + 2.3D - 74.8F + 21.4G \quad \%fit = 92.8$$

Clearly the high inter-correlations cause problems but they are mainly due to the differences in size between regions. Standardising the data by dividing by the number of people will not only help to reduce the inter-correlations but also make the data more meaningful.

2. The intercorrelation matrix is greatly improved.

Population (B)	1.00					
Salespersons (C)	0.05	1.00				
GDP (D)	−0.04	0.14	1.00			
Complaints (E)	0.48	−0.40	0.09	1.00		
Competitors (F)	0.20	−0.08	0.24	0.72	1.00	
Advertising (G)	−0.37	−0.71	−0.27	0.07	−0.14	1.00
	Pop. (B)	**Salesp. (C)**	**GDP (D)**	**Compl. (E)**	**Compet. (F)**	**Advert. (G)**

Only one model comes from this data set.
F (competitors) is entered first (% fit = 82.6, significant at 1% level), then G (advertising) (increase in % fit = 15.0, significant at 1% level).

Constant: 139.46				% fit: 97.6 d.f.: 9	Residual SD: 6.00	
Variables in the equation				Variables available to add		
Variable	Coefficient	Decrease in %fit	Test value to delete	Variable	Increase in %fit	Test value to include
F Compet	−23.0335	71.19	16.42**	B Popl	0.03	0.31
G Adverts	14.5854	14.99	7.54*	C Sperson	0.16	0.76
				D GDP	0.00	0.05
				E Compl	0.02	0.29

The model is:-

$$\text{Sales} = 139 - 23.0F + 14.6G \qquad \%\text{fit} = 97.6$$

i.e: Sales: -23.0 (Competitors) $+ 14.6$ (Advertising)

Both variables are highly significant. However, the most important variable is F, the number of competitors, which explains to some degree the region-to-region variability, but which is not under Kandie's control. The other significant variable, amount spent on advertising, is under Kandie's control.

Chapter 10

(a) Acceptability: the first step shows the following:-

Constant: -14.44			% fit: 97.1	d.f.: 8	Residual SD: 0.43		
Variables in the equation				Variables available to add			
Variable	Coefficient	Decrease in %fit	Test value to delete	Variable	Increase in %fit		Test value to include
A Sugar	2.8445	45.03	11.23**				
B Spice	20.2647	5.65	3.98**				
C ATN	0.2261	0.76	1.46				
A^2	-0.1674	3.30	3.04*				
B^2	-5.4153	4.51	3.56**				
C^2	-0.0027	0.03	0.29				
AB	-0.7860	4.83	3.68**				
AC	0.0183	0.36	1.00				
BC	-0.1470	0.93	1.62				

A, B, A^2, B^2 and AB are all significant. None of the variables involving C is significant so they are removed from the model.

(b) The coefficients for the contour diagram are shown in the left-hand part of the table below:-

Constant: -12.23			% fit: 95.1	d.f.: 12	Residual SD: 0.46	
Variables in the equation				Variables available to add		
Variable	Coefficient	Decrease in %fit	Test value to delete	Variable	Increase in %fit	Test value to include
A Sugar	3.2058	45.57	10.59**	C ATN	0.97	1.65
B Spice	18.5014	6.63	4.04**	C^2	0.03	0.25
A^2	-0.1623	3.27	2.84*	AC	0.17	0.62
B^2	-5.3835	5.30	3.61**	BC	0.75	1.41
AB	-0.8898	6.93	4.13**			

Contour Plot of Acceptability

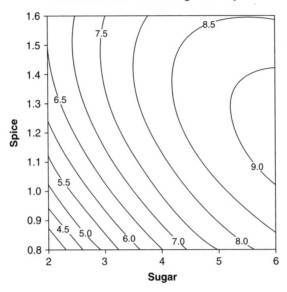

The predicted maximum score of 9.21 occurs at 6% sugar, 1.23 units spice. The contour diagram suggests high scores (8 and above) in a region corresponding to high sugar and a wide range of spice.

(c) Crunchiness: the first step shows the following:-

Constant: 1.37			% fit: 84.1	d.f.: 8		Residual SD: 0.76	
Variables in the equation					Variables available to add		
Variable	Coefficient	Decrease in %fit	Test value to delete	Variable	Increase in %fit		Test value to include
A Sugar	−0.6584	2.19	1.05				
B Spice	13.5524	1.49	0.87				
C ATN	−0.0862	63.28	5.65**				
A^2	0.0821	1.42	0.85				
B^2	−6.3252	11.03	2.36*				
C^2	−0.0076	0.44	0.47				
AB	0.1877	0.49	0.50				
AC	−0.0251	1.20	0.78				
BC	0.1291	1.29	0.81				

From the multiple regression analysis only B^2 and C are significant. None of the variables including A is significant so they are removed from the model.

(d) Only B^2 (and consequently B) and C need to be in the equation, but BC and C^2 will also be kept for completeness.

The coefficients for the contour diagram are shown in the left-hand part of the table below:-.

Constant: −0.01			% fit: 78.6	d.f.: 12	Residual SD: 0.72		
Variables in the equation				**Variables available to add**			
Variable	Coefficient	Decrease in %fit	Test value to delete	Variable	Increase in %fit		Test value to include
B Spice	14.1555	1.12	0.80	A Sugar	2.07		1.09
C ATN	−0.1216	71.63	6.35**	A^2	1.85		1.02
B^2	−6.1358	11.81	2.58*	AB	0.79		0.65
C^2	−0.0092	0.67	0.61	AC	0.81		0.66
BC	0.1037	0.98	0.74				

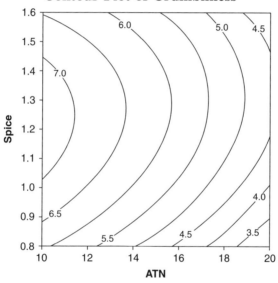

Contour Plot of Crumbliness

The specification for crumbliness is between 4.0 and 7.0, but with 95% confidence intervals for predicted score of about 1.0, we need to look for the region giving scores between 5.0 and 6.0, with any value for spice and values of intermediate ATN depending on the value of spice.

(e) For maximum acceptability we required high sugar and intermediate spice; for crumbliness within specification intermediate ATN and any value of spice.
    Rudyard's should use:

Sugar (A)   6.0
Spice (B)   1.2
ATN (C)   18

Chapter 14

Note: The CAED program uses a random number sequence, so different runs on different computers will give slightly different solutions.

1. (a) The repertoire contains all combinations of the following levels:-

Carbon black (A) :   20, 30, 40

% of natural rubber (B) :   50, 70, 90

Catalyst concentration (C) : 2.0, 3.0

The designs give the following statistics:-

Number of Trials	Det M	Max r
11	0.068	0.39
12	0.078	0.26
13	0.093	0.20
14	0.084	0.20

Det M increases considerably from 11 to 13 trials and Max r decreases. There is considerable gain in using 13 trials but little gain in extending.

(b) The same levels of A and B are used, but C now additionally includes the level of 2.5. The designs give the following statistics:-

Number of Trials	Det M	Max r
11	0.0139	0.39
12	0.0181	0.26
13	0.0277	0.23
14	0.0249	0.35

2. (a) The repertoire used all combinations of Temperature (200, 210, 220, 230, 240, 250, 260, 270, 280, 290 and 300) and Pressure (10, 11, 12, 13, 14, 15 and 16) but excluded those outside the operating region.

Number of Trials	Det M	Max r
14	0.0149	0.40

Max r is high. It involves the correlation between the interaction and pressure. It is due to the unusual design region.

(b) Including duplicates improves Det M to 0.021.
Examining the design matrix we see that three duplicate points were chosen.

Chapter 17

1. (a)

Source of Variation	Sum of Squares	Degrees of Freedom	Mean Square	Test Value
Due to treatments	18.96	5	3.79	5.32
Residual	8.56	12	0.71	
Total	27.52	17		

(b) **Null hypothesis -**   The treatment population means are equal.
    Alternative hypothesis -   The treatment population means are not equal.

    Test value -
$$= \frac{\text{`Due to treatments' mean square}}{\text{Residual mean square}}$$
$$= \frac{3.79}{0.71} = 5.32$$

    Table values -   From Table A.6 (one-sided) with 5 and 12 degrees of freedom:
     3.11 at the 5% significance level;
     5.06 at the 1% significance level.

    Decision -   Reject the null hypothesis at the 1% significance level.
    Conclusion -   The treatment means are significantly different.

(c)

$$\text{Residual SD (s)} = \sqrt{\text{Residual mean square}}$$
$$= 0.84 \text{ based on 12 degrees of freedom}$$
$$5\% \; LSD = ts\sqrt{\frac{1}{n_A} + \frac{1}{n_B}}$$

$s = 0.84$, t (5% significance level, 12 degrees of freedom) $= 2.18$, $n_A = n_B = 3$

$$5\% LSD = 2.18 \times 0.84\sqrt{\frac{1}{3} + \frac{1}{3}}$$
$$= 1.50$$

The means which were not significantly different from each other are bracketed together in the table below:

Treatment	Mean
A	6.8
B	6.2
F	5.8
E	5.4
C	4.4
D	3.8

D appears to be the most effective but is not significantly better than C.

238   SOLUTIONS TO PROBLEMS

2. (a)

Source of Variation	Sum of Squares	Degrees of Freedom	Mean Square
Due to drugs	2.9505	3	0.9835
Residual (within drugs)	1.7587	8	0.2198
Total	4.7092	11	

(b) **Null hypothesis -**   The drug means are equal.
   Alternative hypothesis -   The drug means are not equal.

Test value - $= \dfrac{\text{'Due to drugs' mean square}}{\text{Residual mean square}}$

$= \dfrac{0.9835}{0.2199} = 4.47$

Table values: From Table T6 (one-sided) with 3 and 8 degrees of freedom:
4.07 at the 5% significance level;
7.59 at the 1% significance level.
Decision: We reject the null hypothesis at the 5% significance level.
Conclusion: The drug means are not all equal.

(c)   Residual SD (s) $= \sqrt{\text{Residual mean square}}$
= 0.47 based on 8 degrees of freedom

$$5\% \, LSD = ts\sqrt{\dfrac{1}{n_A} + \dfrac{1}{n_B}}$$

$s = 0.47$, t (5% significance level, 8 degrees of freedom) $= 2.31$, $n_A = n_B = 3$

$$5\% LSD = 2.31 \times 0.47\sqrt{\dfrac{1}{3} + \dfrac{1}{3}}$$
$$= 0.88$$

Means which are not significantly different from each other are bracketed together in the table below:

Drug	Mean
A	3.65
B	4.31
D	4.65
C	4.99

(d) The conclusion is that A is the most effective antibiotic, but is not significantly better than B.

Chapter 18

(a) The days are the "blocks".

(b) Readings were to be taken at different times of the year which would be expected to have different levels in accordance with seasonal weather patterns.

(c)

Source of Variation	Sum of Squares	Degrees of Freedom	Mean Square	Test Value
Between days	738.4	3	246.1	
Between sites	680.0	4	170.0	7.98
Residual	255.6	12	21.3	
Total	1674.0	19		

Table value - From Table A.6 (one-sided) with 4 and 12 degrees of freedom:

$$3.26 \text{ at the } 5\% \text{ significance level;}$$
$$5.41 \text{ at the } 1\% \text{ significance level.}$$

We can conclude that there are differences in SO_2 levels between the sites, at the 1% significance level.

(d) Residual SD (s)=$\sqrt{21.3} = 4.62$ with 12 degrees of freedom.

$$5\% \, LSD = ts\sqrt{\frac{1}{n_A} + \frac{1}{n_B}}$$

$s = 4.62$, t (5% significance level, 12 degrees of freedom) $= 2.18$, $n_A = n_B = 4$

$$5\% LSD = 2.18 \times 4.62\sqrt{\frac{1}{4} + \frac{1}{4}} = 7.1$$

Site	Mean
Kinlochnever	12
Swestport	15
Carville	21
Hardunby	24
Codurham	28

The lowest level was at Kinlochnever, but not significantly less than at Swestport. Either of these sites would be suitable by the stated criterion.

(e) Without the use of the "block" variable, the Residual Sum of Squares would be 994 based on 15 degrees of freedom, giving a Residual Mean Square of 66.3.

The test value would be $170/66.3 = 2.57$

The table value (Table A.6) with 4 and 15 degrees of freedom would be 3.06 at the 5% significance level.

We would be unable to conclude that there were differences in SO_2 levels between the sites.

Chapter 19

1. Degrees of Freedom:

$$
\begin{aligned}
\text{Total} \quad &= 24 - 1 = 23 \\
\text{Resins} \quad &= \text{Number of Resins} - 1 \\
&= 4 - 1 = 3 \\
\text{Position} \quad &= 3 - 1 = 2 \\
\text{Residual} \quad &= 12 \text{ cells each with 1 degree of freedom} \\
&= 12 \\
\text{Interaction} \quad &= 23 - 3 - 2 - 12 \\
&= 6
\end{aligned}
$$

2. Analysis of Variance table:

Source	Sum of Squares	Degrees of Freedom	Mean Square	F	Table Value	P
Between positions	142.5	2	71.2	12.79	3.89	0.001
Between resins	360.2	3	120.1	21.54	3.40	0.000
Interaction	70.4	6	11.7	2.11	3.00	0.128
Residual	66.9	12	5.6			
Total	640.0	23				

3. The analysis of variance table shows that the interaction is not significant. Thus we can test the main effects against the residual and both are shown to be highly significant with very low p-values.

4. The residual SD = 2.36

   t with 12 degrees of freedom = 2.18

   The least significant differences are:

   $$
   \text{For position} = 2.18 \times 2.36 \times \sqrt{\frac{1}{8} + \frac{1}{8}}
   $$
   $$
   = 2.57
   $$

   $$
   \text{For resin} = 2.18 \times 2.36 \times \sqrt{\frac{1}{6} + \frac{1}{6}}
   $$
   $$
   2.97
   $$

Using the means of each treatment gives:

Spray Gun Position	Mean
Bottom	17.75 ⎤
Top	17.16 ⎦
Middle	12.31

Resin	Mean
Present	21.48
Resin C	15.96 ⎤
Resin A	14.86 ⎦
Resin B	10.62

5. The results show that all the new resins are significantly better than the old resin with Resin B being significantly better than both A and C.

There is no significance between top and bottom spray gun positions but the middle position gives significantly lower broken edges. Thus the best combination is Resin B with the middle spray position.

Chapter 20

(a) The number of flavours $= f = 5$
The number of assessors $= b = 10$
The number of scores $\quad = N = 30$

Adjusted flavour mean

$\quad = $ Observed flavour mean

$\quad - \left(\dfrac{fb-N}{f(N-b)} \right)$ (Sum of other flavour means)

$\quad + \left(\dfrac{f-1}{N-b} \right)$ (Sum of other assessor means)

$\quad = $ Observed flavour mean $- 0.2$ (Sum of other flavour means)

$\quad + 0.2$ (Sum of other assessor means)

Flavour	A	B	C	D	E
Flavour mean	6	4	7	2	6
Sum of other flavour means	19	21	18	23	19
Sum of other assessor means	22	21	16	24	17
Adjusted flavour means	6.6	4.0	6.6	2.2	5.6

Before the means are adjusted it appears that flavour C is superior, with flavours A and E in joint second place. However, the apparent popularity of flavour C is partly due to it having been rated by high-scoring assessors. When the flavour means are adjusted we see that A and C are equally desirable.

(b) Total sum of squares

$$= (N-1)(SD \text{ of } N \text{ scores})^2$$
$$= (29)(2.1972)^2$$
$$= 140.0$$

Between assessors sum of squares

$$= (b-1)(SD \text{ of assessor means})^2(No. \text{ of scores per assessor})$$
$$= (9)(1.3333)^2(3)$$
$$= 48.0$$

Adjusted between flavours sum of squares

$$= (N-b)(SD \text{ of corrected fragrance means})^2$$
$$= (20)(1.8921)^2$$
$$= 71.6$$

Source of Variation	Sum of Squares	Degrees of Freedom	Mean Square
Between flavours (adjusted)	71.6	4	17.9
Between assessors (crude)	48.0	9	Not used
Residual	20.4	16	1.275
Total	140.0	29	

(c)

Null hypothesis -	There is no difference in the acceptability of the five flavours.
Alternative hypothesis -	There are differences in the acceptability of the five flavours.

Test value -
$$= \frac{\text{'Due to flavours' mean square}}{\text{Residual mean square}}$$
$$= \frac{17.9}{1.275} = 14.04$$

Table value - From Table A.6 (one-sided) with 4 and 16 degrees of freedom: 3.01 at the 5% level, 4.77 at the 1% level.

Decision - Reject the null hypothesis at the 1% level.

Conclusion - There are differences in the acceptability of the five flavours.

Thus we can confidently conclude that the flavours do differ in their appeal to the users.

(d) Estimate of residual standard deviation
$$= \sqrt{\text{Residual mean square}}$$
$$= \sqrt{1.275}$$
$$= 1.129$$

(e) Least Significant Difference $= t \times (\textit{Residual SD}) \times \sqrt{\dfrac{2(f-1)}{N-b}}$
where

$f =$ number of treatments (flavours)
$b =$ number of blocks (assessors)
$N =$ total number of scores
and t is a value from Table A.2 at a 5% significance level with the same degrees of freedom as the Residual SD.

$$f = 5,\ b = 10,\ N = 30,\ t \text{ with } 16 \text{ degrees of freedom} = 2.12$$

$$5\% \text{ LSD} = 2.12 \times 1.129 \times \sqrt{\frac{2 \times 4}{30-10}} = 1.51$$

Thus any two flavours with a difference of more than 1.51 between their adjusted means can be judged as significantly different.

Flavour	Adjusted Mean
D	2.2
B	4.0
E	5.6
A	6.6
C	6.6

A and C have the highest adjusted means, but not significantly higher than E. They are all significantly more acceptable than B which is more acceptable than D.

Statistical tables

Effective Experimentation: For Scientists and Technologists   Richard Boddy and Gordon Smith
© 2010 John Wiley & Sons, Ltd

Table A.1   Grubbs' test for a single outlier using mean and SD

Degrees of freedom	Significance level	
	5% (0.05)	1% (0.01)
2	1.15	1.15
3	1.48	1.50
4	1.71	1.76
5	1.89	1.97
6	2.02	2.14
7	2.13	2.27
8	2.21	2.39
9	2.29	2.48
10	2.36	2.56
11	2.41	2.64
12	2.46	2.70
13	2.51	2.76
14	2.55	2.81
15	2.59	2.85
16	2.62	2.89
17	2.65	2.93
18	2.68	2.97
19	2.71	3.00
20	2.73	3.03
30	2.92	3.26
40	3.05	3.39
50	3.14	3.49
100	3.38	3.75

Test value $= \dfrac{|x - \bar{x}|}{s}$

where x is the most extreme observation from the mean;

$\bar{x}$ is the mean of all observations including the possible outlier;

s is the SD of all observations including the possible outlier.

Table A.2 The t-table

Degrees of freedom	Significance level					
	10% (0.1)	5% (0.05)	2% (0.02)	1% (0.01)	0.2% (0.002)	0.1% (0.001)
1	6.31	12.71	31.82	63.66	318.31	636.62
2	2.92	4.30	6.97	9.92	22.33	31.60
3	2.35	3.18	4.54	5.84	10.21	12.92
4	2.13	2.78	3.75	4.60	7.17	8.61
5	2.02	2.57	3.37	4.03	5.89	6.87
6	1.94	2.45	3.14	3.71	5.21	5.96
7	1.89	2.36	3.00	3.50	4.79	5.41
8	1.86	2.31	2.90	3.36	4.50	5.04
9	1.83	2.26	2.82	3.25	4.30	4.78
10	1.81	2.23	2.76	3.17	4.14	4.59
11	1.80	2.20	2.72	3.11	4.03	4.44
12	1.78	2.18	2.68	3.06	3.93	4.32
13	1.77	2.16	2.65	3.01	3.85	4.22
14	1.76	2.15	2.62	2.98	3.79	4.14
15	1.75	2.13	2.60	2.95	3.73	4.07
16	1.75	2.12	2.58	2.92	3.69	4.02
17	1.74	2.11	2.57	2.90	3.65	3.97
18	1.73	2.10	2.55	2.88	3.61	3.92
19	1.73	2.09	2.54	2.86	3.58	3.88
20	1.72	2.08	2.53	2.85	3.55	3.85
25	1.71	2.06	2.49	2.78	3.45	3.72
30	1.70	2.04	2.46	2.75	3.39	3.65
40	1.68	2.02	2.42	2.70	3.31	3.55
60	1.67	2.00	2.39	2.66	3.23	3.46
120	1.66	1.98	2.36	2.62	3.16	3.37
Infinity	1.64	1.96	2.33	2.58	3.09	3.29
Confidence level	90%	95%	98%	99%	99.8	99.9%

Note: The quoted significance levels are for a two-sided test. To carry out a one-sided test, halve the significance level given at the top of the table.

Table A.3 Outlier test for two extreme observations, one high and one low

Degrees of freedom	Significance level	
	5% (0.05)	1% (0.01)
4	6.74	15.81
5	3.69	6.55
6	2.92	4.49
7	2.44	3.45
8	2.17	2.84
9	2.00	2.52
10	1.86	2.30
12	1.68	1.99
14	1.57	1.80
16	1.50	1.69
18	1.44	1.61
20	1.39	1.55
30	1.27	1.34
40	1.20	1.27
50	1.17	1.22
100	1.09	1.11

$$\text{Test value} = \frac{\text{SD for all observations}}{\text{SD excluding highest and lowest observations}}$$

Table A.4 Test for two extreme observations, both high or both low

Degrees of freedom	Significance level	
	5% (0.05)	1% (0.01)
4	2.10	2.16
5	2.41	2.50
6	2.66	2.79
7	2.87	3.02
8	3.04	3.22
9	3.18	3.40
10	3.31	3.56
12	3.56	3.82
14	3.75	4.02
16	3.90	4.18
18	4.04	4.33
20	4.17	4.47
30	4.60	4.97
40	4.87	5.32
50	5.08	5.53
100	5.63	6.07

$$\text{Test value} = \frac{|\text{sum of two highest (or lowest) deviations}|}{s}$$

where s is the standard deviation of all observations including the possible outliers.

Table A.5   Outlier test for residuals in simple and multiple regression

Sample size	5% (0.05) significance level									
	No. of coefficients in equation									
	1	2	3	4	5	6	8	10	15	25
5	1.92	1.74								
6	2.07	1.93								
7	2.19	2.08	1.94							
8	2.28	2.20	2.10	1.94						
9	2.35	2.29	2.21	2.10	1.95					
10	2.42	2.37	2.31	2.22	2.11	1.95				
12	2.52	2.49	2.45	2.39	2.33	2.24	1.96			
14	2.61	2.58	2.55	2.51	2.47	2.41	2.25	1.96		
16	2.68	2.66	2.63	2.60	2.57	2.53	2.43	2.26		
18	2.73	2.72	2.70	2.68	2.65	2.62	2.55	2.44		
20	2.78	2.77	2.76	2.74	2.72	2.70	2.64	2.57	2.15	
25	2.89	2.88	2.87	2.86	2.84	2.83	2.80	2.76	2.60	
30	2.96	2.96	2.95	2.94	2.93	2.93	2.90	2.88	2.79	2.17
35	3.03	3.02	3.02	3.01	3.00	3.00	2.93	2.97	2.91	2.64
40	3.08	3.08	3.07	3.07	3.06	3.06	3.05	3.03	3.00	2.84
45	3.13	3.12	3.12	3.12	3.11	3.11	3.10	3.09	3.06	2.96
50	3.17	3.16	3.16	3.16	3.15	3.15	3.14	3.14	3.11	3.04
60	3.23	3.23	3.23	3.23	3.22	3.22	3.22	3.21	3.20	3.15
70	3.29	3.29	3.28	3.28	3.28	3.28	3.27	3.27	3.26	3.23
80	3.33	3.33	3.33	3.33	3.33	3.33	3.33	3.32	3.31	3.29
90	3.37	3.37	3.37	3.38	3.38	3.38	3.36	3.37	3.36	3.34
100	3.41	3.41	3.40	3.40	3.40	3.40	3.40	3.40	3.39	3.38

↑
Simple regression
$y = a + bx$

For simple regression :

$$\text{Test value} = \frac{\text{Maximum}|\text{Residual}|}{\text{RSD}\sqrt{1 - \dfrac{1}{n} - \dfrac{(X-\bar{x})^2}{(n-1)(\text{SD of } x)^2}}}$$

Table A.6 The F-test (two-sided)

5% (0.05) significance level

Degrees of freedom for smaller SD	Degrees of freedom for larger SD														
	1	2	3	4	5	6	7	8	9	10	12	15	20	60	Infinity
1	647.8	799.5	864.2	899.6	921.8	937.1	948.2	956.7	963.3	968.6	976.7	984.9	993.1	1010	1018
2	38.51	39.00	39.17	39.25	39.30	39.33	39.36	39.37	39.39	39.40	39.41	39.43	39.45	39.48	39.50
3	17.44	16.04	15.44	15.10	14.88	14.73	14.62	14.54	14.47	14.42	14.34	14.25	14.17	13.99	13.90
4	12.22	10.65	9.98	9.60	9.36	9.20	9.07	8.98	8.90	8.84	8.75	8.66	8.56	8.36	8.26
5	10.01	8.43	7.76	7.39	7.15	6.98	6.85	6.76	6.68	6.62	6.52	6.43	6.33	6.12	6.02
6	8.81	7.26	6.60	6.23	5.99	5.82	5.70	5.60	5.52	5.46	5.37	5.27	5.17	4.96	4.85
7	8.07	6.54	5.89	5.52	5.29	5.12	4.99	4.90	4.82	4.76	4.67	4.57	4.47	4.25	4.14
8	7.57	6.06	5.42	5.05	4.82	4.65	4.53	4.43	4.36	4.30	4.20	4.10	4.00	3.78	3.67
9	7.21	5.71	5.08	4.72	4.48	4.32	4.20	4.10	4.03	3.96	3.87	3.77	3.67	3.45	3.33
10	6.94	5.46	4.83	4.47	4.24	4.07	3.95	3.85	3.78	3.72	3.62	3.52	3.42	3.20	3.08
12	6.55	5.10	4.47	4.12	3.89	3.73	3.61	3.51	3.44	3.37	3.28	3.18	3.07	2.85	2.72
15	6.20	4.77	4.15	3.80	3.58	3.41	3.29	3.20	3.12	3.06	2.96	2.86	2.76	2.52	2.40
20	5.87	4.46	3.86	3.51	3.29	3.13	3.01	2.91	2.84	2.77	2.68	2.57	2.46	2.22	2.09
60	5.29	3.93	3.34	3.01	2.79	2.63	2.51	2.41	2.33	2.27	2.17	2.06	1.94	1.67	1.48
Infinity	5.02	3.69	3.12	2.79	2.57	2.41	2.29	2.19	2.11	2.05	1.94	1.83	1.71	1.39	1.00

Table A.6 *(Continued)*

1% (0.01) significance level

Degrees of freedom for smaller SD	Degrees of freedom for larger SD														
	1	2	3	4	5	6	7	8	9	10	12	15	20	60	Infinity
1	16211	20000	21615	22500	23056	23437	23715	23925	24091	24224	24426	24630	24836	25253	25465
2	198.5	199.0	199.2	199.2	199.3	199.3	199.4	199.4	199.4	199.4	199.4	199.4	199.4	199.5	199.5
3	55.55	49.80	47.47	46.19	45.39	44.84	44.43	44.13	43.88	43.69	43.29	43.08	42.78	42.15	41.83
4	31.33	26.28	24.26	23.15	22.46	21.97	21.62	21.35	21.14	20.97	20.70	20.04	20.17	19.61	19.32
5	22.78	18.31	16.53	15.56	14.94	14.51	14.20	13.96	13.77	13.62	13.38	13.15	12.90	12.40	12.14
6	18.63	14.54	12.92	12.03	11.46	11.07	10.79	10.57	10.39	10.25	10.03	9.81	9.59	9.12	8.88
7	16.24	12.40	10.88	10.05	9.52	9.16	8.89	8.68	8.51	8.38	8.18	7.97	7.75	7.31	7.08
8	14.69	11.04	9.60	8.81	8.30	7.95	7.69	7.50	7.34	7.21	7.01	6.81	6.61	6.18	5.95
9	13.61	10.11	8.72	7.96	7.47	7.13	6.88	6.69	6.54	6.42	6.23	6.03	5.83	5.41	5.19
10	12.83	9.43	8.08	7.34	6.87	6.54	6.30	6.12	5.97	5.85	5.66	5.47	5.27	4.86	4.64
12	11.75	8.51	7.23	6.52	6.07	5.76	5.52	5.35	5.20	5.09	4.91	4.72	4.53	4.12	3.90
15	10.80	7.70	6.48	5.80	5.37	5.07	4.85	4.67	4.54	4.42	4.25	4.07	3.88	3.48	3.26
20	9.94	6.99	5.82	5.17	4.76	4.47	4.26	4.09	3.96	3.85	3.68	3.50	3.32	2.92	2.69
60	8.49	5.79	4.73	4.14	3.76	3.49	3.29	3.13	3.01	2.90	2.74	2.57	2.39	1.96	1.69
Infinity	7.88	5.30	4.28	3.72	3.35	3.09	2.90	2.74	2.62	2.52	2.36	2.19	2.00	1.53	1.00

Test value $= \dfrac{\text{Larger } s^2}{\text{Smaller } s^2}$

Table A.6 The F-test (one-sided) for use in analysis of variance

5% (0.05) significance level

Degrees of freedom for residual mean square	Degrees of freedom for treatment mean square														
	1	2	3	4	5	6	7	8	9	10	12	15	20	60	Infinity
1	161.4	199.5	215.7	224.6	230.2	234.0	236.8	238.9	240.5	241.9	243.9	246.0	248.0	252.2	254.3
2	18.51	19.00	19.16	19.25	19.30	19.33	19.35	19.37	19.38	19.40	19.41	19.43	19.45	19.48	19.50
3	10.13	9.55	9.28	9.12	9.01	8.94	8.89	8.85	8.81	8.79	8.74	8.70	8.66	8.57	8.53
4	7.71	6.94	6.59	6.39	6.26	6.16	6.09	6.04	6.00	5.96	5.91	5.86	5.80	5.69	5.63
5	6.61	5.79	5.41	5.19	5.05	4.95	4.88	4.82	4.77	4.74	4.68	4.62	4.56	4.43	4.36
6	5.99	5.14	4.76	4.53	4.39	4.28	4.21	4.15	4.10	4.06	4.00	3.94	3.87	3.74	3.67
7	5.59	4.74	4.35	4.12	3.97	3.87	3.79	3.73	3.68	3.64	3.57	3.51	3.44	3.30	3.23
8	5.32	4.46	4.07	3.84	3.69	3.58	3.50	3.44	3.39	3.35	3.28	3.22	3.15	3.01	2.93
9	5.12	4.26	3.86	3.63	3.48	3.37	3.29	3.23	3.18	3.14	3.07	3.01	2.94	2.79	2.71
10	4.96	4.10	3.71	3.48	3.33	3.22	3.14	3.07	3.02	2.98	2.91	2.85	2.77	2.62	2.54
12	4.75	3.89	3.49	3.26	3.11	3.00	2.91	2.85	2.80	2.75	2.69	2.62	2.54	2.38	2.30
15	4.54	3.68	3.29	3.06	2.90	2.79	2.71	2.64	2.59	2.54	2.48	2.40	2.33	2.16	2.07
20	4.35	3.49	3.10	2.87	2.71	2.60	2.49	2.45	2.39	2.35	2.28	2.20	2.12	1.95	1.84
60	4.00	3.15	2.76	2.53	2.37	2.25	2.17	2.10	2.04	1.99	1.92	1.84	1.75	1.53	1.39
Infinity	3.84	3.00	2.60	2.37	2.21	2.10	2.01	1.94	1.88	1.83	1.75	1.67	1.57	1.32	1.00

Table A.6 *(Continued)*

1% (0.01) significance level

Degrees of freedom for residual mean square	Degrees of freedom for treatment mean square														
	1	2	3	4	5	6	7	8	9	10	12	15	20	60	Infinity
1	4052	5000	5403	5625	5764	5859	5928	5982	6022	6056	6106	6157	6209	6313	6366
2	98.50	99.00	99.17	99.25	99.30	99.33	99.36	99.37	99.39	99.40	99.42	99.43	99.45	99.48	99.50
3	34.12	30.82	29.46	28.71	28.24	27.91	27.67	27.49	27.35	27.23	27.05	26.87	26.69	26.32	26.13
4	21.20	18.00	16.69	15.98	15.52	15.21	14.98	14.80	14.66	14.55	14.37	14.20	14.02	13.65	13.46
5	16.26	13.27	12.06	11.39	10.97	10.67	10.46	10.29	10.16	10.05	9.89	9.72	9.55	9.20	9.02
6	13.75	10.92	9.78	9.15	8.75	8.47	8.26	8.10	7.98	7.87	7.72	7.56	7.40	7.06	6.88
7	12.25	9.55	8.45	7.85	7.46	7.19	6.99	6.84	6.72	6.62	6.47	6.31	6.16	5.82	5.65
8	11.26	8.65	7.59	7.01	6.63	6.37	6.18	6.03	5.91	5.81	5.67	5.52	5.36	5.03	4.86
9	10.56	8.02	6.99	6.42	6.06	5.80	5.61	5.47	5.35	5.26	5.11	4.96	4.81	4.48	4.31
10	10.04	7.56	6.55	5.99	5.64	5.39	5.20	5.06	4.94	4.85	4.71	4.56	4.41	4.08	3.91
12	9.33	6.93	5.95	5.41	5.06	4.82	4.64	4.50	4.39	4.30	4.16	4.01	3.86	3.54	3.36
15	8.68	6.36	5.42	4.89	4.56	4.32	4.14	4.00	3.89	3.80	3.67	3.52	3.37	3.05	2.87
20	8.10	5.85	4.94	4.43	4.10	3.87	3.70	3.56	3.46	3.37	3.23	3.09	2.94	2.61	2.42
60	7.08	4.98	4.13	3.65	3.34	3.12	2.95	2.82	2.72	2.63	2.50	2.35	2.20	1.84	1.60
Infinity	6.63	4.61	3.78	3.32	3.02	2.80	2.64	2.51	2.41	2.32	2.18	2.04	1.88	1.47	1.00

$$\text{Test value} = \frac{\text{Effect } MS}{\text{Residual } MS}$$

Table A.7 Cochran's test for a high outlier amongst standard deviations

Number of groups	Number of observations per group									
	2		3		4		5		6	
Sig. level	5% (0.05)	1% (0.01)	5% (0.05)	1% (0.01)	5% (0.05)	1% (0.01)	5% (0.05)	1% (0.01)	5% (0.05)	1% (0.01)
2	–	–	0.98	1.00	0.94	0.98	0.91	0.96	0.88	0.94
3	0.97	0.99	0.87	0.94	0.80	0.88	0.75	0.83	0.71	0.79
4	0.91	0.97	0.77	0.86	0.68	0.78	0.63	0.72	0.59	0.68
5	0.84	0.93	0.68	0.79	0.60	0.70	0.54	0.63	0.51	0.59
6	0.78	0.88	0.62	0.72	0.53	0.63	0.48	0.56	0.45	0.52
7	0.73	0.84	0.56	0.66	0.48	0.57	0.43	0.51	0.40	0.47
8	0.68	0.79	0.52	0.62	0.44	0.52	0.39	0.46	0.36	0.42
9	0.64	0.75	0.48	0.57	0.40	0.48	0.36	0.43	0.33	0.39
10	0.60	0.72	0.45	0.54	0.37	0.45	0.33	0.39	0.30	0.36
11	0.57	0.68	0.42	0.50	0.35	0.42	0.31	0.37	0.28	0.33
12	0.54	0.65	0.39	0.48	0.33	0.39	0.29	0.34	0.26	0.31
13	0.52	0.62	0.37	0.45	0.31	0.37	0.27	0.32	0.25	0.29
14	0.49	0.60	0.35	0.43	0.29	0.35	0.26	0.30	0.23	0.27
15	0.47	0.58	0.34	0.41	0.28	0.33	0.24	0.29	0.22	0.26
16	0.45	0.55	0.32	0.39	0.26	0.32	0.23	0.27	0.21	0.25
17	0.43	0.53	0.31	0.37	0.25	0.30	0.22	0.26	0.20	0.23
18	0.42	0.51	0.29	0.36	0.24	0.29	0.21	0.25	0.19	0.22
19	0.40	0.50	0.28	0.34	0.23	0.28	0.20	0.24	0.18	0.21
20	0.39	0.48	0.27	0.33	0.22	0.27	0.19	0.23	0.17	0.21
25	0.33	0.41	0.25	0.31	0.19	0.22	0.16	0.19	0.14	0.17
30	0.29	0.36	0.20	0.24	0.16	0.19	0.14	0.16	0.12	0.15
35	0.26	0.33	0.18	0.21	0.14	0.17	0.12	0.14	0.11	0.13
40	0.24	0.29	0.16	0.19	0.13	0.15	0.11	0.13	0.10	0.11

Test value $= \dfrac{\text{largest } s^2}{\sum s^2}$

Table A.8 Analysis matrices for 2^2, 2^3 and 2^4 factorial experiments

Trial	A	B	AB	C	AC	BC	ABC	D	AD	BD	ABD	CD	ACD	BCD	ABCD
1	−	−	+	−	+	+	−	−	+	+	−	+	−	−	+
2	+	−	−	−	−	+	+	−	−	+	+	+	+	−	−
3	−	+	−	−	+	−	+	−	+	−	+	+	−	+	−
4	+	+	+	−	−	−	−	−	−	−	−	+	+	+	+
5	−	−	+	+	−	−	+	−	+	+	−	−	+	+	−
6	+	−	−	+	+	−	−	−	−	+	+	−	−	+	+
7	−	+	−	+	−	+	−	−	+	−	+	−	+	−	+
8	+	+	+	+	+	+	+	−	−	−	−	−	−	−	−
9	−	−	+	−	+	+	−	+	−	−	+	−	+	+	−
10	+	−	−	−	−	+	+	+	+	−	−	−	−	+	+
11	−	+	−	−	+	−	+	+	−	+	−	−	+	−	+
12	+	+	+	−	−	−	−	+	+	+	+	−	−	−	−
13	−	−	+	+	−	−	+	+	−	−	+	+	−	−	+
14	+	−	−	+	+	−	−	+	+	−	−	+	+	−	−
15	−	+	−	+	−	+	−	+	−	+	−	+	−	+	−
16	+	+	+	+	+	+	+	+	+	+	+	+	+	+	+

Table A.9 Rank order statistics for half-normal plots

8 Trials		12 Trials		16 Trials	
Order	Value	Order	Value	Order	Value
1	0.1597	1	0.1054	1	0.0787
2	0.3260	2	0.2128	2	0.1583
3	0.5042	3	0.3234	3	0.2393
4	0.7021	4	0.4388	4	0.3221
5	0.9344	5	0.5609	5	0.4075
6	1.2349	6	0.6926	6	0.4962
7	1.7239	7	0.8381	7	0.5893
		8	1.0047	8	0.6879
		9	1.2057	9	0.7941
		10	1.4730	10	0.9102
		11	1.9215	11	1.0404
				12	1.1914
				13	1.3763
				14	1.6256
				15	2.0509

Index

Effective Experimentation: For Scientists and Technologists   Richard Boddy and Gordon Smith
© 2010 John Wiley & Sons, Ltd